MATHEMATICS ASSESSMENT
AND
EVALUATION

SUNY Series, Reform in Mathematics Education
Judith Sowder, editor

The preparation of this book was supported by the Office for Educational Research and Improvement, United States Department of Education (Grant Number R117G10002) and by the Wisconsin Center for Education Research, School of Education, University of Wisconsin-Madison. The opinions expressed in this publication do not necessarily reflect the views of the Office of Educational Research and Improvement or the Wisconsin Center for Education Research.

MATHEMATICS ASSESSMENT
AND
EVALUATION

Imperatives for Mathematics Educators

Edited by

Thomas A. Romberg

STATE UNIVERSITY OF NEW YORK PRESS

Published by
State Univesity of New York Press, Albany

© 1992 State University of New York

For information, address the State University of New York Press,
State University Plaza, Albany, NY 12246

Production by Christine M. Lynch
Marketing by Fran Keneston

Library of Congress Cataloging-in-Publication Data

Mathematics assessment and evaluation : imperatives for
  mathematics educators / edited by Thomas A. Romberg.
      p.   cm. — (SUNY series, reform in mathematics
  education)
    Includes bibliographical references and index.
    ISBN 0-7914-0899-X (CH : acid-free). — ISBN 0-7914-
  0900-7 (PB : acid-free)
      1. Mathematical ability—Testing. 2. Mathematics—
  Study and teaching—United States. I. Romberg, Thomas
  A. II. Series. 510'.71—dc20   91-11157
                                                                CIP

10 9 8 7 6 5 4 3 2 1

CONTENTS

1

Overview of the Book

Thomas A. Romberg

The purpose of this book is to share with mathematics educators a set of recent papers written on issues surrounding mathematics tests and their influence on school mathematics. The impetus for the contributions grew from a conference, "The Influence of Testing on Mathematics Education," sponsored by the Mathematical Sciences Education Board (MSEB) at UCLA in June 1986. The purpose of the conference was to gather informed input and advice on current testing practices. The fact is that students in American schools are subjected to a variety of tests, often standardized tests, from kindergarten to graduate school. Such tests are, according to widely held perception, inhibitors to change and improvement in education and especially in mathematics education. Since MSEB was organized to coordinate the current reform movement in school mathematics, the topic of the conference was deemed critical. Two things became clear at the UCLA meeting: First, there was agreement that tests need to change to reflect curriculum changes, and second, many participants articulated their beliefs about the inadequacy of current tests and provided relevant anecdotes on problems to others at the conference. However, no one was sure how such changes could be accomplished, nor did participants even have substantive, reliable information about the actual impact of testing on classroom practices.

Following the conference, a three-person Testing Design Task Force, consisting of Jeremy Kilpatrick, Tej Pandey, and Thomas Romberg, was organized by MSEB. In September 1986, this task force produced "A Proposal for Studies of Mathematics Tests and Testing" based on the following four assumptions:

- Valid information about student mathematics is needed by a variety of people (students, teachers, parents, administrators, policy makers) for a variety of purposes (monitoring progress, selection for and placement in courses, program evaluation, accountability).

- Both the curriculum and teaching practice in mathematics need to be directed toward strategies which students could use to solve problems, the application of mathematics to practical situations, and the development of thinking skills. Consequently, testing should reflect students' achievement in these directions.

- Serious questions have arisen about the validity of existing tests for the uses to which they are being put. Standardized tests and state-mandated tests may yield information that is invalid for certain purposes and provide little or no information on several important dimensions of achievement.

- The continued use of existing tests appears likely to impede the much-needed reform in curriculum and instruction to which the mathematics education community is committed.

On the basis of these assumptions, a set of questions and research studies was proposed. In particular, several literature reviews were planned, each of which would explore one facet of the validity of mathematics tests for various purposes. Topics were to include surveys of testing practice, the alignment of tests with curricula, test-preparation practices and effects, test-taking skills, the student use of calculators during test taking, teacher and student attitudes toward tests, time spent in testing, alternatives to testing, and minority group and gender group differences in risk taking and test performance.

In 1987, when the Wisconsin Center for Education Research was awarded the grant to form the National Center for Research in Mathematical Sciences Education (NCRMSE), NCRMSE assumed responsibility for carrying out several aspects of the proposed scope of work outlined by the MSEB task force. The papers in this volume represent a number of the literature reviews proposed. They were written by the Center staff or by invited scholars. The contributions cover many of the issues identified by the MSEB task force and are an important contribution to our growing knowledge about the impact of tests and testing on school mathematics.

The papers are only a part of the work now being conducted by the Center on this important topic. Since 1987, the Center has conducted two major surveys. The first, a national survey of a sample of Grade 8 mathematics teachers (Romberg, Zarinnia, & Williams, 1989), provides information about teachers' perceptions of the impact of mandated testing on their instruction. Findings reveal that teachers are familiar with mandated tests, make efforts to ensure that students perform well on the tests, and adjust their curriculum and modes of instruction to focus on the knowledge and skills being tested.

The second is a survey of state mathematics coordinators on the current types of mandated testing in the fifty states (Romberg, Zarinnia, & Williams, 1990). This study examines the actual mandated testing practices in each state, including the kinds of tests given, the uses to which they are put, and the kinds of test-score information subsequently available to the teachers.

In addition to these surveys and this collection of papers, four related activities are now in progress:

1) During the past year, two in-depth case studies on the impact of mandated testing in classrooms have been conducted at four sites. Information from these studies is now being analyzed.

2) Three extensive reviews of literature and of practice are now underway on classroom testing for instructional decision making, testing for placement and grouping, and test validity.

3) Some sample test items have been written and are being tried out; they have been designed to assess level of reasoning in some of the particular domains outlined in the *Curriculum and Evaluation*

Standards for School Mathematics (National Council of Teachers of Mathematics, 1989).

4)   Two projects on curriculum design, development, and assessment are being conducted jointly with the Research Group in Mathematics Education at the University of Utrecht.

In all, the work of the staff and consultants of NCRMSE on the influence of testing on school mathematics makes it clear that valid information about student performance is sorely needed if the reform movement in school mathematics is to succeed.

OVERVIEW OF THE CHAPTERS

The next twelve chapters were prepared during 1988 and 1989. Chapter 2 represents a summary of the overall problems associated with the need for valid information. Chapters 3 and 4 examine the use of tests in the context of the current reform movement in school mathematics. Chapters 5 and 6 describe the current procedures used to develop state tests. Chapter 7 summarizes current efforts to incorporate the use of calculators in mathematics tests. This is followed by chapter 8, a review of research on testing with calculators. Chapter 9 is a review of gender differences and testing. Chapter 10 is an examination of an Australian project addressing teachers' assessment practices. The next two chapters, 11 and 12, deal with alternative strategies for gathering, analyzing, and reporting student performance information. The final chapter is an invited review and critique of chapters 2 through 12.

Chapter 2: Evaluation: A Coat of Many Colors
by Thomas A. Romberg

An earlier draft of this chapter was prepared as an invited address for Theme Group—T4, Evaluation and Assessment, at the Sixth International Congress on Mathematical Education in Hungary. This paper examines both the methods of gathering information from students and the use of that information to make a variety of judgments. It considers the history of evaluation and how evaluation relates to the gathering of assessment data and to educational decision making. To examine the strengths and weaknesses of the evaluation of the impact of new mathematics programs and of large-scale profile evaluations, it describes trends in evaluation and assessment that show the disparity between what is possible and what is, in fact, achieved.

Chapter 3: Implications of the NCTM *Standards* for Mathematics Assessment by Norman Webb and Thomas A. Romberg

In 1987 work began on the NCTM *Standards*. Thomas Romberg was the chair of the commission that produced this document and Norman Webb chaired the working group that prepared the evaluation standards. This chapter includes criteria for assessment which would be compatible with and supportive of the curriculum standards. Three examples of alternative assessment techniques are presented that correspond to the intent of the evaluation standards and provide illustrations of forms of assessment that are applicable in evaluating the curriculum standards.

Chapter 4: Curriculum and Test Alignment by Thomas A. Romberg, Linda Wilson, 'Mamphono Khaketla, and Silvia Chavarria

In this chapter, a variety of tests and test items are examined to determine whether they reflect the recommendations made in the *Standards*. In the initial sections, six commonly used standardized tests are examined. It is clear from this examination that those tests fail to assess the higher-order skills such as problem solving, reasoning, and connections that are stressed in the *Standards*. Then items are identified from other tests which could be used to assess such aspects of mathematics.

Chapter 5: State Assessment Test Development Procedures by James Braswell

The primary purpose of this paper is to describe how tests are developed for state assessment programs. The methods described are based in part on discussions with state department of education assessment staff members in Florida, Louisiana, Massachusetts, Michigan, and New Jersey—states in which testing practice was judged to be representative of a range of approaches to test development. Occasional references that reflect previous experience with other state testing programs and current work with the National Assessment of Educational Progress test development team for the 1990 Mathematics Assessment are also taken into consideration.

Chapter 6: Test Development Profile of a State-Mandated Large-Scale Assessment Instrument in Mathematics
by Tej Pandey

Two main types of large-scale assessments are examined in this paper. The first focus of interest is oriented primarily to those individuals who typically use test information to rank a student on an established norm, find a student's strengths and weaknesses, and determine whether that student has mastered specific course content. The second focus of interest lies primarily in the administrative use of information to determine the achievement level of students in a school, district, or regional system for purposes of assessing program effectiveness. This paper examines the nature and design of test instruments in a large-scale assessment program (California Assessment Program) providing reliable group-level information. The paper also describes the test development process as it has evolved over a period of fifteen years to meet the curriculum demands of the time.

Chapter 7: Assessing Students' Learning in Courses Using Graphics Tools: A Preliminary Research Agenda
by Sharon L. Senk

Recently mathematics educators have called for the use of calculator and computer-graphing technology in mathematics classes, and several software and curriculum development projects have been initiated to transform these recommendations into reality. However, until now, there has been little systematic study of how teaching, learning, and assessment in courses using such graphics tools are affected by the technology. This paper describes a preliminary agenda developed by researchers in the field for assessing students' learning in courses using graphing tools. Included are suggested investigations of student and teacher outcomes and a discussion of methodological issues.

Chapter 8: Mathematics Testing with Calculators: Ransoming the Hostages
by John G. Harvey

This paper argues that present testing practices hold today's students hostage to yesterday's mistakes. The author predicts that because mathematics tests fail to incorporate the use of calculators in the testing process, mathematics instruction will fail to incorporate the use of calculators effectively, continuing to hold today's students prisoners to a mathematics curriculum that is failing to prepare them for society's immediate needs as well as those of the twenty-first century. The paper suggests that the use of calculators on mathematics tests will not remedy the failures of present tests, but that their use is necessary if we want students to investigate, to explore, and to discover mathematics effectively.

Chapter 9: Gender Differences in Test Taking: A Review
by Margaret R. Meyer

Ideally, when students take a mathematics examination, the only thing that should influence their score is their mastery of the material being tested. This paper reviews evidence concerning the existence of gender differences in mathematics test taking. It examines several factors that have surfaced relating to differences in performances for males and females. These factors are power vs. speed test conditions, item-difficulty sequencing, examination format, test-wiseness, risk-taking behavior, and test-preparation behaviors. One conclusion reached is that the use of the multiple-choice format could result in a male advantage. A recommendation is therefore made that assessment instruments not rely as heavily on the multiple-choice format.

Chapter 10: Communication and the Learning of Mathematics
by David Clarke, Max Stephens, and Andrew Waywood

The learning of mathematics is fundamentally a matter of constructing mathematical meaning. The environment of the mathematics classroom provides experiences which stimulate this process of construction. This chapter presents the findings of three studies based in Australian schools: the IMPACT Project, Assessment Alternatives in Mathematics, and the Vaucluse College Study. The purpose of the research synthesis considered in this chapter is to discuss (a) the extent to which the strategies reported encourage children to broaden their mathematical thinking and facilitate meta-learning and (b) the impact of these strategies on the nature of mathematical activity in classrooms, in particular with reference to redefining the roles of teacher and student in creating and giving personal meaning to mathematics.

Chapter 11: Measuring Levels of Mathematical Understanding
by Mark Wilson

This chapter describes recent psychometric advances in the creation of models that measure developmental change in understanding. Standardized, norm-referenced tests are based on an accumulation of bits of knowledge rather than on understanding, which is a constructivist, developmental process. As the latter conception gains more acceptance, there is a need for new assessment models. Empirical examples of response maps are used to illustrate the potential of the new models.

Chapter 12: A Framework for the California Assessment Program to Report Students' Achievement in Mathematics
by E. Anne Zarinnia and Thomas A. Romberg

The purpose of this paper is to propose categories for the California Assessment Program that report student achievement in mathematics. Initially, the purpose of reporting achievement was accountability. This paper examines explicit and tacit messages imposed in the analyzing, gathering, and aggregating of this information that expose subtle effects on teaching and student achievement. The paper determines that units of analysis and reporting categories are needed that will both deliberately support the purposes of gathering adequate information for monitoring and—by focusing attention on critical considerations—promote reform in mathematics education. This chapter outlines seven bases for forming reporting categories.

Chapter 13: Evaluation—Some Other Perspectives
by Philip C. Clarkson

A common response to the challenge of the *Standards* is, "Yes, but who will change the tests?" (National Council of Teachers of Mathematics, 1989, p. 189).

It is apparent that "the tests" referred to are not the tests teachers give in their classrooms on a day-to-day or weekly basis. They have control over those already. "The tests" are the standardized assessment instruments which are used throughout the United States, often authorized by legislation, devised by commercial organizations, and seen by many teachers in the country as being a forceful factor in structuring their mathematics curriculum.

The preceding chapters have provided background information on these tests and have made some suggestions on how they could be altered. None reflect on the question as to whether they are indeed necessary. This chapter sketches developments over the last twenty-five years in the State of Victoria, Australia, where there is now only one external test given at the end of the school system, in Year 12. This contrasting situation may contribute constructively to the ongoing debate in both Australia and in the United States as to how to monitor the work of schools.

In summary, as the title to this book suggests, the authors of these chapters address an important set of issues about mathematics assessment and evaluation. It is clear that it is important to gather information on student performance in mathematics for a variety of reasons. However, while the mathematics curriculum and the way mathematics is taught are changing, the definition of assessment and how performance is assessed also need to change. It is imperative, if the school mathematics reform efforts are to be successful, that mathematics educators become aware of the issues addressed in these chapters.

2

Evaluation: A Coat of Many Colors

Thomas A. Romberg

"EVALUATE: to judge or determine the worth or quality of."
Webster's New World Dictionary, 1985.

This paper examines both the methods of gathering information from students and the use of that information to make a variety of judgments. It considers the history of evaluation, and how evaluation relates to the gathering of assessment data and to educational decision making. To examine the strengths and weaknesses of evaluations of the impact of new mathematics programs and large-scale profile evaluations, it describes trends in evaluation and assessment, showing the disparity between what is possible and what is already being done.

Evaluation in education has evolved from an initial and single concentration on the measurement of achievement in order to make judgments about students to the current and growing interest in providing information to support policy and program decision making. To make these latter judgments, information from students about their mathematical achievement is usually used. Thus, in this paper both the methods of gathering information from students and the uses of that information to make a variety of judgments are examined.

The assessment of student performance in schools has a long history. However, contemporary models for the gathering of performance data and the use of the information for policy and program decision making have only evolved during the past quarter-century. The purposes of this survey paper are

1. To relate the gathering of assessment data to educational decision making;

2. To trace the history of this evolution. The assessment history begins in the nineteenth century and the evaluation history in the 1930s. However, against the background of each case, developments in the past decade are stressed;

3. To illustrate the strengths and weaknesses of two contemporary social policy evaluation models. These are: evaluations of the impact of new mathematics programs, and large-scale profile evaluations; and

4. To describe four recent trends in evaluation and assessment.

Although the history of and trends in assessment and evaluation are not unique to school mathematics, the emphasis and examples in this paper focus on assessing mathematical performance and on the use of that information in instructional and policy contexts. Also, the examples have been selected to reflect the variety in models, methods, and procedures used throughout the world.

The principal point which should be understood is that at present there is considerable disparity between theory and practice. Academic considerations about goals, decisions, methods of gathering information, and the validity of that information are in sharp contrast to the political and practical expectations of many governments and administrators. What is possible differs from what is done.

EDUCATIONAL DECISION MAKING

The following examples are provided to illustrate the relationships between measures of achievement and the variety of situations in which this information is used to make a judgment (hence, the title of this chapter):

1)   A student has decided to study biology and would like to know whether she has the prerequisite knowledge to enroll in a biometrics course.

2)   The admissions committee of a tertiary institution must select one hundred students from some eight hundred who have applied for an engineering program.

3)   A teacher would like to grade students on how well they understand the chapter on simultaneous linear equations just completed.

4)   An official in a department of education has been asked to provide a legislative committee with information about pupil performance in mathematics.

5)   A publishing company is interested in developing a text to teach specific concepts of statistics to students in middle school. It needs feedback from teachers about the adequacy of the materials (i.e., what things were successful and what things were not) so that improvements can be made.

6)   A researcher interested in early cognitive development with respect to mathematics would like to assess the ability of preschool children to handle certain mathematical relationships, such as the comparison of two sets with respect to numerosity.

7)   An employer is interested in the mathematical capability of job applicants.

8)   An official must decide which students are to be admitted to academic high schools and which to technical schools.

These examples are only a few of the typical situations in which information from students about their mathematical performance is frequently used. In addition, they reflect the diversity of judgment (qualification, selection, placement, diagnosis, grading, profiling, researching, and so forth) involved in those decisions as well as the variety of personnel (students, administrators, teachers, developers, employers, and researchers).

From these examples, I have assumed that information from students about their mathematical achievement is important and that such information should influence educational decisions. The scenarios cited here are but a few examples of the many decisions facing educators throughout the world. Whether achievement data as a source of information actually influence schooling decisions is a separate and distinct empirical question. Nevertheless, valid data about student achievement should be available and used when making many such decisions.

Also, we must ask: How should such information be elicited? The answer to this question will be based on a second assumption: The methods of gathering information (how data is collected, from whom, and how it is aggregated, organized, and reported) depend on the decisions to be made.

From these assumptions and the examples given above, I believe three elements of the decision-making process should be considered.

1) The decisions must be specifically identified. Gathering information without an explicit purpose in mind wastes time and resources. Although it is now fashionable to create data bases under the assumption that having such data will be useful, it has been shown that such data bases are rarely used or of value unless the purposes for which the data are to be used were considered when designing the data base.

2) The implications of the judgments to be made, or the questions to be answered, must be examined. This involves considering error in measurement, the errors in judgment that one is willing to tolerate, and whether the decisions are irrevocable. Teachers may be willing to accept considerable measurement error when administering chapter tests because they can rely on other information to judge a student's progress; a developer may be willing to live with high judgment errors in the development of a new instructional unit, while an admissions committee should seek minimal measurement error in choosing which applicants to accept into a program.

3) The "unit" about which the decisions are to be made must be specified (individuals, groups, classes, schools, materials, research questions). It has long been common practice to test all students on every item in every test; data from individuals can then be aggregated at any group level

for any purpose. This practice is extremely wasteful, in terms of both cost and time. For example,
the administration of a standardized test merely
to publish the results in the local press (as is
common in the United States) is wasteful both of
student time and district resources—that is, the
cost of administration and scoring. Profiling school
performance can be accomplished more efficiently
by other means.

In summary, to assess student performance in mathematics, one should consider the kinds of judgments or evaluations
that need to be made and tailor the assessment procedures to
the decisions that will be made on the basis of those judgments. This is particularly important when the information is
being used by policy makers to make programmatic decisions.

HISTORY OF ASSESSMENT AND EVALUATION

The history of the measurement of human behavior, with primary reference to the capacities and educational attainments
of school students, may be divided roughly into four periods.
During the first period, from the beginning of the historical
record to the nineteenth century, measurement in education
was quite crude. During the second period, embracing approximately the whole of the nineteenth century, educational measurement began to assimilate, from various sources, the ideas
and the scientific and statistical techniques which were later to
result in the psychometric testing movement. The third period,
dating from about 1900 to the 1960s, can be characterized as
the psychometric period. The final period, dating from the 1960s
to the present, is the policy-program evaluation period.

Early Examination

The initiation ceremonies by which primitive tribes tested the
knowledge of tribal customs, endurance, and the readiness of
the young for admission to the ranks of adulthood may be
among the earliest examinations employed by human beings.
Use of a crude oral test was reported in the Old Testament, and
Socrates is known to have employed searching types of oral
quizzing. Elaborate and exhaustive written examinations were

used by the Chinese as early as 2200 B.C. in the selection of their public officials. These illustrations may be classified as historical antecedents of performance tests, oral examinations, and essay tests. However, there is no evidence that different individuals ever took the same tests and all judgments were made by officials in a manner similar to that used in examining doctoral students today.

Educational Testing in the Nineteenth Century

Three persons made outstanding contributions to educational testing in the nineteenth century. The ideas of these men— Horace Mann, George Fisher, and J. M. Rice—set the precedent for developments during the present century.

The first school examinations of note appear to have been instituted in the United States, in the Boston schools in 1845, as substitutes for oral tests when enrollments became so large that the school committee could no longer examine all pupils orally. These written examinations, in arithmetic, astronomy, geography, grammar, history, and natural philosophy, impressed Horace Mann, then secretary of the Massachusetts Board of Education. As editor of the *Common School Journal*, he published extracts from them and concluded that the new written examination was superior to the old oral test in these respects:

1.   It is impartial.
2.   It is just to the pupils.
3.   It is more thorough than older forms of examination.
4.   It prevents the "officious interference" of the teacher.
5.   It "determines, beyond appeal or gainsaying, whether the pupils have been faithfully and competently taught."
6.   It removes "all possibility of favoritism."
7.   It makes the information obtained available to all.
8.   It enables public appraisal of the ease or difficulty of the questions. (Greene, Jorgenson, & Gerberich, 1953)

Although these ideas are similar to the objectives represented by modern tests, the instruments themselves were inadequate. However, in successive issues of the *Common School*

Journal, Mann was to suggest most of the principles of examination that are found in contemporary measurement—for example, timed responses by students to identical series questions.

To the Reverend George Fisher, an English schoolmaster, goes the credit for devising and using what were probably the first objective measures of achievement. His "scale books," used in the Greenwich Hospital School as early as 1864, provided the means for evaluating accomplishments in handwriting, spelling, mathematics, grammar and composition, and in several other school subjects. Specimens of pupil work were compared with "standard specimens" to determine numerical ratings that, at least for spelling and a few other subjects, depended on errors in performance (Greene, Jorgenson, & Gerberich, 1953). Scoring procedures for many examinations still follow this procedure (e.g., the English O level examinations).

The use of test information for program evaluation was first developed by J. M. Rice, an American dentist. In 1894, he developed a battery spelling test. Having administered a list of spelling words to pupils in many school systems and analyzed the results, Rice found that pupils who had studied spelling thirty minutes a day for eight years were no better spellers than children who had studied the subject fifteen minutes a day for eight years. Rice was attacked and reviled for this "heresy," and some educators even attacked the use of a measure of how well pupils could spell as a means of evaluating the efficiency of spelling instruction. They intended that spelling be taught to develop the pupils' minds and not to teach them to spell. It was more than a decade later that Rice's pioneering effort resulted in significant attention to objective models in educational testing (Ayres, 1918).

The Psychometric Period

This era began shortly after the turn of the century. Although the historical antecedents sketched in the preceding paragraphs were essential prerequisites, developments first in mental testing and shortly after in achievement testing lay at the roots of testing progress in this era.

General Intelligence Tests. Attempts to measure general intelligence, ability to learn, and ability to adapt oneself to new situations had been made both in the United States and in

France. The first individual test was developed in France, and the first group test was developed some years later in the United States.

Individual intelligence scales were originated in 1905 by Binet and Simon in France. Their first scale was devised primarily for the purpose of selecting mentally retarded pupils who required special instruction. This pioneer individual-intelligence scale was based on interpreting the relative intelligence of different children at any given chronological age by the number of questions they could answer of varied types and increasing levels of difficulty. These characteristics were all reembodied in the 1908 and 1911 revisions of the Binet-Simon Scale and remain basic to most individual intelligence scales today. The 1908 revision introduced the fundamentally important concept of mental age (MA) and provided a means for determining it (Freeman, 1930).

The first group intelligence test was *Army Alpha*, used for the measurement and placement of American army recruits and draftees during World War I. It was the product of the collaboration of various psychologists working on group intelligence tests when the United States entered the war.

Aptitude Tests. The measurement of aptitudes, or those potentialities for success in an area of performance that exist prior to direct acquaintance with that area, was closely related to intelligence testing. Early attempts to measure general intelligence tested many specific traits and aptitudes, but this approach was abandoned after Binet showed that tests of more complex forms of behavior were superior. It was soon apparent, however, that general intelligence tests were not highly predictive of certain types of performance, especially in the trades and industry. Munsterberg's aptitude tests for telephone girls and streetcar motormen were followed by tests of mechanical aptitude, musical aptitude, art aptitude, clerical aptitude, and aptitude for various subjects of the high school and college curriculum (Watson, 1938). Spearman's (1904) splitting of total mental ability into a general factor and many specific factors had a decided influence on this movement.

Achievement Tests. Modern achievement testing was stimulated by Thorndike's (1904) book on mental, social, and educational measurements. Through his book and his influence on his students, Thorndike was predominantly responsible for the early development of standardized tests. Stone, a student of Thorndike's, published the first arithmetic reasoning test in 1908. Between 1909 and 1915, a series of arithmetic tests and scales for measuring abilities in English composition, spelling, drawing, and handwriting were published (Odell, 1930). During the more than half century since these early testing efforts, literally thousands of standardized achievement tests have been published.

The reasons for presenting this brief history of testing are threefold. First, what is referred to as the "modern testing movement" began with a selection problem (Binet & Simon) and a placement problem (*Army Alpha*). It was assumed that a single measure (e.g., MA) or index (e.g., IQ) could be developed to compare individuals on what was assumed to be a general, fixed, unidimensional trait. In turn, the procedures that evolved in developing and administering these tests were used in aptitude and achievement tests. Second, the testing procedures now considered typical in many countries were developed for group administration of early intelligence tests. Such tests comprise a set of questions (items), each having one unambiguous answer. In this sense, the tests are "objective," since no allowance is made for subjective inferences. Third, subjects are administered the same items under standard (nearly identical) conditions with the same instructions, time, and constraints. Furthermore, subjects' answers can be easily scored as correct or not, the total number of correct answers tallied, tallies transformed, and transformed scores compared. Psychometrics, involving the application of statistical procedures to such tests, developed as a field of study in the 1920s.

Most importantly, it should be understood that the testing movement was a product of a historical era. It grew out of the machine-age thinking of the industrial revolution of the last century. Business, industry, and, in particular, schools have been conceived, modified, and operated based on this mechanical view of the world since before the turn of the century.

The Policy-Program Evaluation Period

Information about student achievement has long been used by teachers and educators to make decisions about students. However, the use of that information for wide-scale policy or program judgments is recent. It began with the burst of reform policies associated with the mid-sixties Great Society initiatives in the United States. Federal-level insistence on evaluation of these initiatives was thrust upon a largely unprepared field. In areas as diverse as bilingual education, career education, compensatory programs, reading, or mathematics, little expertise in evaluation existed in the very agencies responsible for carrying out program evaluation. In fact, in the United States the initial training institute on program evaluation was held at the University of Illinois in 1963 (directed by Lee Cronbach).

That early work followed the notions of Ralph Tyler (1931), the "father of educational evaluation." His conception of evaluation involved comparison between intended and observed program objectives. Tyler's model of evaluation in education prevailed until the 1970s, when his approach, like traditional social science models, was found inadequate as a guide for policy and practice. The Tyler evaluation model was based on the hypothetico-deductive traditions of "hard science." It focused on outcomes and sought significant differences between intended and observed outcomes. Initial evaluations of federal education programs used experimental methodologies to assess student achievement and program effectiveness. As applied, this approach paid little attention to the context of program activities or the processes by which program plans were translated into practice (Eash, 1985; O'Keefe, 1984). The discourse about evaluation included fairly rigid rules for "good" design and "scientific" evaluation. In particular, evaluators gathered data on student performance using standard achievement tests.

In summary, evaluation for policy and program purposes began in the 1960s by attempting to apply "scientific" principles that used concepts from the experimental sciences. The information on students was from tests based on the psychometric assessment technique outlined above. Again, this approach to evaluation is a product of "industrial age" thinking.

TWO SOCIAL POLICY EVALUATION MODELS

Policy makers (legislators, government officials, school administrators, and other educators) must make many decisions related to the teaching and learning of mathematics. In this section, two evaluation models often used by policy makers are examined in detail so that their strengths and weaknesses become apparent.

Program Evaluation

Attempts to evaluate the impact of a new curriculum program involved comparison of the performance of a group of students who had studied mathematics from that curriculum with an alternate group, most often a nonequivalent group. Performance was measured in both groups based on scores derived from the same instrument. Initially, in the United States standardized tests were used; later it became common to use objective-referenced tests—that is, tests that produced scores related to specific instructional objectives.

Norm-referenced standardized tests have become an annual ritual in most American schools. Such tests are designed to indicate a respondent's position in a population. Each test comprised a set of independent, multiple-choice questions. The items have necessarily been subjected to a preliminary trial with a representative pupil group so that it is possible to arrange them in the desired manner with respect to difficulty and the degree to which they discriminate among students. Also, the test is accompanied by a chart or table to be used to transform test results into meaningful characterizations of pupil mental ability or achievement (grade-equivalent scores, percentiles, stanines).

Three features of such tests merit comment. First, although each test is designed to order individuals on a single (unidimensional) trait, such as quantitative aptitude, the derived score is not a direct measure of that trait. Second, because individual scores are compared with those of a norm population, there will always be some high and some low scores. This is true even if the range of scores is small. Thus, high and low scores cannot fairly or accurately be judged as "good" or "bad" with respect to the underlying trait. Third, test items are assumed to be equivalent to one another. They are selected on the basis of general level of difficulty (p-value) and some index of discrimination (e.g., nonspurious biserial correlation). Furthermore, the test items are not representative of any well-defined domain.

The primary strengths of standardized tests are that they are relatively easy to develop, inexpensive, and convenient to administer. Furthermore, the results are readily comprehensible since standard procedures are followed.

Their primary weakness is that they are often used as a basis for decisions they were not designed to address. For example, aggregating standardized scores for students in a class (or in a school, or district) to produce a class profile of achievement (class mean) is very inefficient. The tests provide too little information in light of the high cost involved. In fact, it has become clear that such tests are of little value for most evaluations, since the items are not intended to be representative of the mathematical domains in the curriculum.

Unfortunately, in the United States their use appears to be more strongly related to political, rather than educational, uses. For example, it is claimed that elected officials and educational administrators increasingly use the scores from such tests in comparative ways—to indicate which schools, school districts, and even individual teachers give the appearance of achieving better results (National Coalition of Advocates for Students, 1985). Such comparisons are simply misleading. One can only conclude that standardized tests are unwisely overused.

Objective-referenced tests are a product of the behavioral objectives movement in the 1960s. They were developed to provide teachers with an objective set of procedures with which to make instructional decisions. Item development was based on the identification of a set of behavioral objectives such as, "The subject, when exposed to the conditions described in the antecedent, displays the action specified in the verb in the situation specified by the consequent to some specified criterion" (Romberg, 1976, p. 23). Items randomly selected from a pool designed to represent the antecedent conditions and the same action verb are given to students. From their responses, diagnosis of problems or judgments of mastery of objectives can be made.

Three features of these tests should be mentioned. First, they usually are designed as part of a curriculum and meant to be administered to individuals at the end of a specific instructional topic. Often, they are given individually, and teachers' judgments are made quickly. Second, they have occasionally been used in group settings. For example, the comprehensive achievement monitoring scheme (Gorth, Schriber, & O'Reilly, 1974) periodically assesses student performance on a set of objectives. Third, decisions about performance are made with respect to certain a priori criteria.

The strengths of objective-referenced tests lie in their usefulness in instruction. As long as instruction on a topic focuses on the acquisition of some specific concept or skill, such tests can be used to indicate whether or not the concept has been learned or the skill mastered. Furthermore, such tests are scored easily and are readily interpretable.

Objective-referenced tests have four weaknesses. First, specifying a set of behavioral objectives fractionates mathematical knowledge. In no way is it possible to reflect in these tests the interrelatedness of concepts and procedures in any domain. Second, objective-referenced tests are costly to construct, because hundreds of objectives are included in any instructional program. Third, simple aggregation across objectives is not reasonable, since objectives are interdependent. Fourth, and most importantly, items for higher-level or complex problem-solving processes are very difficult to construct and are usually omitted. In fact, as used, these tests reinforce the factory metaphor of schooling. They clearly do not reflect how students reason about problem situations, interpret results, or build arguments.

The problem faced by most program evaluators in the 1960s was a direct result of the "scientific" tradition. The only evidence deemed of value was student performance at the end of treatment when compared with that of an alternate treatment group, and the evidence was gathered from either a standardized test or, later, a criterion-referenced test. The results of examinations (such as the National Longitudinal Study of Mathematical Abilities, Begle & Wilson, 1970) did not show that the new program was uniformly superior to the old program, but rather that different curricula were associated with different patterns of achievement.

Policy Profiles

Profile tests are intended to provide information on a variety of mathematical topics so that policy makers can compare individuals or groups in terms of those topics. Profile tests have become very popular. They have been developed for several major studies of mathematical performance, including the National Assessment of Educational Progress (NAEP) in the United States, the First International Mathematics Study (FIMS), the Second International Mathematics Study (SIMS), and the Assessment Performance Unit (APU) in England.

Five features of profile assessments distinguish them from previous tests. First, they make no assumption of an underlying single trait; rather, the tests are designed to reflect the multidimensional nature of mathematical content. Second, items similar to those used in standardized or objective-referenced tests are used. However, it must be acknowledged that the mathematics profiles developed by the APU in England (Foxman et al., 1980, 1981) differ from most other profile assessments in the choice and form of items or exercises administered. Their exercises include a variety of open-ended questions, performance tasks, and other items. Third, the unit of investigation is a group rather than an individual. Matrix sampling is usually used so that a wider variety of items can be included. Fourth, comparisons between groups are shown graphically on actual scores so that no transformations are needed (see, for example, Figures 2-1 and 2-2). Finally, validity is determined

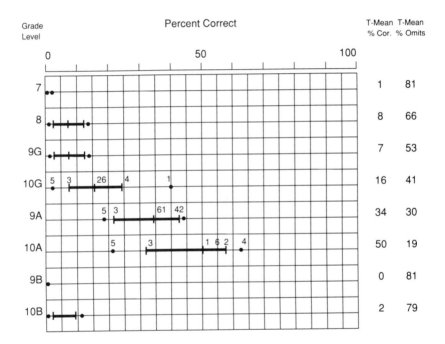

— this topic is not part of the Grade 7 or Grade 8 program.

— a surprisingly large number of Grade 10 Advanced students omitted these items.

— results indicate that where this skill is needed in Grade 11 and 12 it should be reviewed and practiced then.

Figure 2–1. Algebra—equations and inequalities. Range of correct responses to the six instruments, by grade (from McLean, 1982, p. 207).

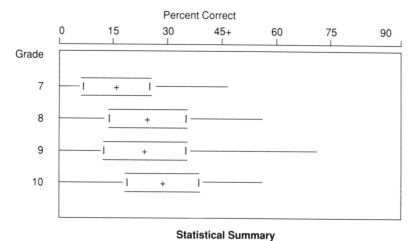

Figure 2–2. Range of correct responses to topic group by grade given in percentages. (McLean, 1982, p. 138).

in terms of content and/or curriculum. Mathematicians and teachers are asked to judge whether individual items reflect a content-by-behavior cell in a matrix. In fact, the usual approach in profile testing is to specify a content-by-behavior matrix. For example, to establish a framework for an item domain, a content-by-behavior grid was developed for each target population in SIMS (Weinzweig & Wilson, 1977). The content dimensions for both Grade 8 and Grade 12 populations were intended to cover all topics likely to be taught in any country. For Grade 8, the content outline contained 133 categories under five broad classifications: arithmetic, algebra, geometry, statistics, and measurement. For Grade 12, the content description was broader, containing 150 categories under seven headings: sets and relations, number systems, algebra, geometry, elementary fractions and calculus, probability and statistics, and finite mathematics.

For each population in SIMS, the behavior dimension referred to four levels of cognitive complexity expected of students: computation, comprehension, application, and analysis. This classification is adapted from Bloom's *Taxonomy of Educational Objectives* (1956). The adaptation involved replacing "knowledge" with "computation" and eliminating the higher levels of synthesis and evaluation. Data from such tests can then be reported in several ways. First, they can be reported in terms of items or cell means. For example, in Figure 2–1, the means are presented for six items on a topic (each given a different instrument) for different students at different grades in the province of Ontario, Canada (McLean, 1982). Second, item scores can be aggregated by columns to yield cognitive level scores or by rows to yield topic scores (see Figure 2–2).

One strength of profile achievement tests is that they can provide useful information about groups; thus they are particularly useful for evaluating educational policy changes that directly affect classroom instruction. However, profile achievement tests are weak in four specific areas. First, because they are designed to reflect group performance, they are not useful for ranking or diagnosing individuals. An individual student takes only a sample of items. Second, they are somewhat more costly to develop and harder to administer and score than standardized norm-referenced tests. Third, because they yield a profile of scores, they are often difficult to interpret. Finally, however, the primary weakness of most profile achievement tests centers on the outdated assumptions underlying the two dimensions of content-by-behavior matrices. The content dimension involves a classification of mathematical topics into "informational" categories. As I have argued:

> "Informational knowledge" is material that can be fallen back upon as given, settled, established, assured in a doubtful situation. Clearly, the concepts and processes from some branches of mathematics should be known by all students. The emphasis of instruction, however, should be "knowing how" rather than "knowing what." (Romberg, 1983, p. 122)

Furthermore, items in any content category are tested as if they were independent of one another, a practice that ignores the interconnections between ideas within a well-defined mathematical domain. Schoenfeld and Herrmann (1982) cautioned about the problems inherent in testing students on isolated tasks.

> If they succeed on those problems, we and they con-
> gratulate each other on the fact that they have learned
> some powerful mathematical techniques. In fact, they
> may be able to use such techniques mechanically while
> lacking some rudimentary thinking skills. To allow them,
> and ourselves, to believe that they understand the math-
> ematics is deceptive and fraudulent. (p. 29)

Thus, the items should reflect the interdependence, rather than independence, of ideas in a content domain.

The behavior dimension of matrices has always posed problems. All agree that Bloom's *Taxonomy* (1956) has proven useful for low-level behaviors (knowledge, comprehension, and application), but difficult for higher levels (analysis, synthesis,
and evaluation). Single-answer, multiple-choice items are not
reasonable at higher levels. One problem is that the *Taxonomy*
suggests that "lower" skills should be taught before "higher"
skills. The fundamental problem is the *Taxonomy's* failure to
reflect current psychological thinking on cognition, and the fact
that it is based on "the naive psychological principle that individual simple behaviors become integrated to form a more complex behavior" (Collis, 1987, p. 3). In the past thirty years, our
knowledge about learning and about how information is processed has changed and expanded.

Bloom's taxonomy of educational objectives epitomizes the
domination of American education by scientific management,
for it completed the process by which not only the content of
learning, but the proxies for its intelligent application, were
classified, organized in a linear sequence and, by definition,
broken into a hierarchy of mutually exclusive cells. The consequences in the classroom were far reaching. Scope and sequence charts prescribed which parts of a subject were to be
covered in what order; each cellular part of each subject was
put into a matrix (e.g., Romberg & Kilpatrick, 1969, p. 285);
behaviors suggesting desirable intellectual activity were also
sequenced. However, given the multiplicity of subject cells to be
covered, the easiest way to finish the prescribed course of study
was to simply cover content without worrying too much about

thought. Furthermore, matrices are difficult to construct effectively on paper in more than two dimensions. Consequently, few scope and sequence charts addressed, in a coherent manner, both levels of thinking and specific aspects of content within an overall discipline. Thus, one focus of concern in documents addressing the quality of education has been the failure of students to attain "higher order intellectual skills" (National Commission on Excellence in Education, 1983, p. 9).

Attacking behaviorism as the bane of school mathematics, Eisenberg (1975) criticized the dubious merit of a task-analysis, engineering approach to curricula, because it essentially equates training with education, missing the heart and essence of mathematics. Expressing concern over the validity of learning hierarchies, he argued for a reevaluation of the objectives of school mathematics. The goal of school mathematics is to teach students to think, to make them comfortable with problem solving, to help them question and formulate hypotheses, investigate, and simply tinker with mathematics. In other words, the focus is turned inward to cognitive mechanisms.

I believe that instruments for assessment should embody the following commonalities:

1. All knowledge is rooted in experience.
2. Knowledge entails the structural modeling of perceived regularities and the reconciling of irregularities.
3. Cohesion of structure is integral and derived from purpose.
4. Quality is determined by predictive power.
5. Disequilibrium is essential to the process.
6. Knowledge is both individual and communal.

Simply stated, there is a need for tools that document the production of knowledge and not merely the proxies that contribute to the process, such as time spent learning or the quality of the teaching staff. A sufficiently detailed view of the process is essential in order to have some idea of how to construct policies for intervention. However, if there is any lesson to be learned from the old paradigm, it is that parts of the process cannot be analyzed in isolation, and then aggregated, with the result regarded as an adequate indicator.

In summary, profiling is important, but current profile tests fail to reflect the way mathematical knowledge is structured or how information is processed within mathematical domains.

In this section four trends are described. The first three are academic or theoretical trends apparent in the literature on assessment and evaluation. The last is a conservative political and practical trend which, in some respects, runs counter to the other trends.

The Trend in Program Evaluation

Far from the limited alternatives of "treatment/control" or randomized designs (see Campbell & Stanley, 1966), contemporary evaluators have developed a diverse assortment of evaluation approaches from which to choose—given purpose, context, and program stage. In contrast to the "one right way" approach of the 1960s, today evaluators have multiple (and not always compatible) approaches. This trend began in the 1970s when scholars trained in disciplines other than experimental psychology were asked to assist in educational evaluations. Scholars like Michael Young (1975), Michael Apple (1979), and a little later Thomas Popkewitz (1984), who were trained in anthropology, sociology, and political science respectively, brought the methods of information gathering and analysis of those disciplines to evaluation. In fact, the list of names of designations for the new methods and models can be confusing to someone unfamiliar with the field of evaluation and the controversies that underlie the various empirical procedures. For example, the catalogue of choices now available to evaluators includes: goal-free evaluation (Scriven, 1974); advocate evaluation (Stake & Gjerde, 1974; Reinhard, 1972); connoisseurship (Eisner, 1976); user-driven evaluation (Patton, 1980); ethnographic evaluation (Fetterman, 1984); responsive evaluation (Stake, 1974); naturalistic inquiry (Guba & Lincoln, 1981).

These diverse approaches to evaluation differ in many respects. Chief among them are the role of the evaluator (from educator to management consultant to assessor to advocate), the role of the client (from active stakeholder and collaborator to passive recipient of the evaluation product), the overall design (from experimental or quasi-experimental to exploratory), and focus (on process—formative evaluation; or outcome—summative evaluation). Each of these dimensions corresponds to the contingencies upon which evaluation choices are based: purpose, decision context, stage of program development, status of theory or knowledge base. One consequence for product development was the specification of four stages of evaluation: (1) product design stage—developing a needs assessment; (2) product creation stage—gathering formative data to improve the product; (3) product implementation stage—demonstrating differences between products and making sure appropriate support services are available; and (4) product illuminative stage—an in-depth examination of how the product is actually used (Romberg, 1975).

Another consequence has been the use of a convergent strategy, that is, using several different evaluation models with the same program. For example, in the IGE (Individually Guided Education) Evaluation Study which I directed (Romberg, 1985), we gathered data about reading and mathematics in our school sites in four phases. Phase 1 involved large-scale survey procedures (including the use of a standardized test). Phase 2 was a follow-up study examining the validity of the Phase 1 data. Phase 3 was an ethnographic study of six exemplary IGE schools. Finally, Phase 4 was a detailed examination in Grades 2 and 5 using time-on-task observations and the repeated administration of criterion-referenced tests.

Note also that evaluation experts began calling for better and different instrumentation to gather information about student performance. Overall, while program evaluation models have proliferated and the questions which they must address have become clear, the information used to answer questions too often still comes from inappropriate tests.

It is only recently that it has become apparent that the kind of evidence one needs to gather to judge many programs is, of necessity, different from that obtained from conventional assessment procedures. Tests given in a restricted format (e.g., multiple-choice items) and in a restricted time fail to capture the most important aspects of doing mathematics. During the past decade researchers have developed a plethora of procedures for gathering information from students: think-aloud interview procedures, performance tasks, projects (both individual

and group), hierarchical reasoning tasks, and others. Unfortunately, with one notable exception, these procedures have not
been used in program evaluations because of cost of administration.

The exception is the evaluation of the Hewet Mathematics A
Project in The Netherlands (de Lange, 1987). In that evaluation
five different tasks were used to gather information: timed written tests, two-stage tasks, a take-home task, an essay task,
and an oral task. The overall picture of how well students
learned, developed from information from the five tasks, is much
more enriched than would have been the case if the researchers had used any one task.

Trends in External Assessment

While past assessment procedures are useful for some purposes and undoubtedly will continue to be used, they are products of an earlier era in educational thought. Like the Model T
Ford assembly line, objective tests were considered in the 1920s
as an example of the application of modern scientific techniques. Today, we are both technologically and intellectually
equipped to improve on outdated methods and instruments.
The real problem is that all three forms of tests (profile, standardized, and criterion-referenced) are based on the same set of
assumptions: an essentialist view of knowledge, a behavioral
theory of learning, and a dispensary approach to teaching. It
should be obvious that new assessment techniques need to be
developed which are consistent with a different view of knowledge, learning, and teaching.

New evaluation models are being developed which demand
new assessment procedures. One approach is based on the
specification of mathematical domains and the development of
items that reflect that domain (Romberg, 1987). In turn, this
assessment approach grows out of the extensive research on
such domains. A good example is the work of Gerard Vergnaud
with respect to "conceptual fields" (cf. 1982). The principles
that are followed in this approach include:

> *Principle 1.* A set of specific and important mathematical
> domains needs to be identified, and the structure and
> interconnectedness of the procedures, concepts, and
> problem situations in each of the domains needs to be
> specified.

Note that this approach is different from the current approach to specifying the mathematical content of a test in that networks are being defined rather than categories. This means that the interconnections of concepts and procedures with problem situations are as important as mastery of any node (e.g., a specific procedure). For example, consider the following exercise in second grade addition and subtraction:

Sue received a box of candy for her birthday. She shared twenty-seven pieces with her friends and now has thirty-seven pieces left. How many pieces of candy were originally in the box?

To solve this exercise, a child would be expected first to represent the quantitative information with the subtraction sentence [ ] – 27 = 37. Second, the sentence should be transformed to the addition sentence 27 + 37 = [ ]; then the addition should be performed to yield an answer. What is important is that the child must know that separating situations can be represented by subtraction sentences, that subtraction sentences can be transformed into equivalent addition sentences, and that there are procedures for performing additions. Each piece of knowledge, while important, contributes to a solution process or way of reasoning about a situation that is more important than any single concept or process.

Principle 2. A variety of tasks should be constructed that reflect the typical procedures, concepts, and problem situations of the chosen mathematical domain.

This is a key principle in that the envisioned tasks are not just a more clever set of paper-and-pencil, multiple-choice test items. Although some typical test items may be appropriate for determining mastery of a specific concept or process, many of the tasks must be different. For example, some should be exercises that require the student to relate several concepts and procedures, such as those in the example, from the addition and subtraction given above. (Note: See also the discussion of the Journey problem and Figures 3–1 and 3–2 in the next chapter.) Other tasks may emphasize the level of reasoning associated with a set of questions about the same situation such as the superitem (in Figure 2–3). Still other tasks may ask students to carry out a physical process, such as gather data,

This is a machine that changes numbers. It adds the number you put in three times and then adds 2 more. So, if you put in 4, it puts out 14.

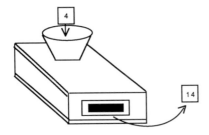

U. If 14 is put out, what number was put in?

Answer: 4

Answer _____

Comment: Students have to understand the problem well enough to be able to close on the correct response which is displayed in the stem.

M. If we put in a 5, what number will the machine put out?

Answer: 17

Answer _____

Comment: Students need to comprehend the set problem sufficiently to be able to use the given statements as a recipe and thus perform a sequence of closures which they do not necessarily relate to one another.

R. If we got out a 41, what number was put in?

Answer: 13

Answer _____

Comment: An integrated understanding of the statements in the problem is necessary to carry out a successful solution strategy in this case. Correct solutions may involve working backwards or carrying out a series of approximation trials. It should be noted that the solution requires only data-constrained reasoning in that no abstract principles need to be invoked.

E. If "X" is the number that comes out of the machine when the number "Y" is put in, write down a formula which will give us the value of "Y" whatever the value of "X."

Answer _____

Answer: $Y = \dfrac{X - 2}{3}$

Comment: A correct response involves extracting the relationships from the problem and setting them down in an abstract formula. It involves using the information given in a way quite different from that of the lower levels.

Figure 2–3. An example of a super item (Collis, Romberg, & Jurdak, 1986, p. 12).

measure an object, construct a figure, work in a group, to organize a simulation. And still others may be open ended, like the Roller Coaster problem (in Figure 2–4).

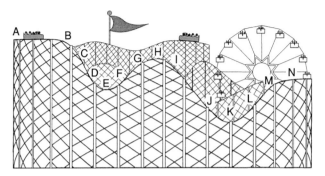

The picture above shows the track of a free-wheeling roller-coaster, which is travelling at a walking pace between A and B.

1. Write a paragraph describing how you think the speed of the roller coaster varies as it travels along the track. (Use the letters A to O to help you in your description.)

2. Now sketch a graph which shows how the speed varies as it travels along the track. (Don't expect to get it right the first time!)

Figure 2–4. Interpreting a roller-coaster (Swan, 1986, p. 36).

These illustrations demonstrate that there are several different aspects of doing mathematics within any mathematical domain. To be able to assess the level of maturity an individual or group has achieved in a domain requires that a rich set of tasks be constructed.

> *Principle 3.* Some tasks in a particular domain would be administered to students via tailored testing, and for groups via matrix sampling as well.

Not all tasks for a domain need to be given to a student or group to determine the level of maturity. The technology is available to systematically vary several aspects of any exercise or problem situation. For example, for the subtraction exercise under Principle 1, one could vary the situations (join-separate, part-part-whole, comparison, etc.), the size of the numbers, the transformations, and the computational strategies (counting, algorithms, and others).

> *Principle 4.* Based on the tasks administered to a student in a domain, their complexity, and the student's responses to those tasks, the information should be logically combined to yield a score for that domain.

Note that this score is not just the number of the correct answers a student has found. Instead, it would involve Boolean combinations of information (such as, following inferential rules like "If _____ and _____, then _____"). The intent of the score is that it reflect the degree of maturity the student has achieved with respect to that domain. Note that this assumes all students are capable of some knowledge in several domains.

> *Principle 5.* A score vector over the appropriate mathematical domains would be constructed for each individual or group. Thus, for any individual one would have several scores $(x_1, x_2, \ldots, x_n)$ where x_i is the score for a particular domain.

Note that this simply reinforces the notion that mathematics is a plural noun.

In summary, awareness of a problem, such as the need for alternative testing procedures for school mathematics, does not mean solutions are easy. It may take years to replace current testing procedures in schools. Nevertheless, this should not deter us from exploring plausible alternatives. What is needed are tasks that provide students an opportunity to reflect, organize, model, represent, and argue within specific domains. Constructing, scoring, scaling, and interpreting responses to such tasks for mathematical domains will not be easy, but will, in the long run, be well worth the effort.

Trends in Assessment by Teachers

One striking consequence of scientific, psychometric assessment procedures has been to deskill teachers. External objective assessment was deemed better than professional judgment. Today, too many teachers are no longer trained in evaluation and lack confidence in their ability to judge student performance (Apple, 1979). In reaction to this awareness, a trend to empower teachers is emerging. For example, the Graded Assessment Project in England (Close & Brown, 1988) provides teachers with procedures to assess performance. This theme is central to the North American National Council of Teachers of Mathematics' *Curriculum and Evaluation Standards* (1989). It is also a major component in the Australian Mathematics Curriculum Teaching Project (MCTP) (Clarke, 1987) and a focal part of the Cognitively Guided Instruction research project at the University of Wisconsin (Peterson, Fennema, Carpenter, & Loef, 1987).

Political-Practical Trend

In most of the world, it is generally agreed that the educational system, as a whole, and the teaching of the learning of mathematics, in particular, need to change. Demands are being made of governments, politicians, and administrators for funds to bring about reform. In turn, of course, these groups have a right to demand that evidence be gathered to prove that their monies are well spent, that changes are in fact made, and that the changes make a difference. Valid pupil performance data are the kinds of information demanded.

However, governmental expectations about such data in the United States and Great Britain revert back to the scientific-experimental notions of the past: behavioral objectives, norm-referenced scores, Bloom's *Taxonomy*. For example, *attainment targets* in the new national curriculum in Great Britain is merely a new label for behavioral objectives. The use of SIMS items for policy profiles (e.g., in Italy and in some parts of the United States) continues the practice of not assessing problem-solving strategies, communication skills, level of reasoning, and other vital areas. These, along with other examples, make it clear that there is considerable disparity between current theory and these practical demands. The demands for information are legitimate. The validity of procedures is suspect.

CONCLUSIONS

The field of assessment and evaluation has come a long way during the last quarter century. However, a lot of work still needs to be done. Assessment of growth in specific domains has replaced general assessment of status performance over several domains.

Unless we make changes in the way in which we gather information from students, we will only contribute to the ongoing difficulties of sterile lessons and the further deskilling of teachers, and we will lose a major opportunity to change the way mathematics is done. Instead, we need to conceive of curricular evaluations and of assessments of individual progress in light of mathematical maturity in specific domains.

1.   Current testing procedures are unlikely to provide valid information for decisions about the current reform movement.

Current tests reflect the ideas and technology of a different era and world view. They cannot assess how students think or reflect on tasks, nor can they measure interrelationships of ideas.

2.   Work should be initiated, or extended, to develop new assessment procedures.

Only by having new assessment tools that reflect authentic achievement in specific mathematical domains can we provide educators with appropriate information about how students are performing. Of necessity, this implies that considerable funds be allocated for research and development. Only when new instruments are developed will we no longer be bound by old assessment procedures rooted in the traditions of the industrial age.

3.   The emerging variety of evaluation models needs to utilize assessment procedures that reflect the changes in school mathematics.

Today, school mathematics is changing the emphasis from drill on basic mathematical concepts and skills to explorations that teach students to think critically, to reason, to solve problems. The criteria for judging level of performance by a student or group of students should be based on these notions. This will involve the student's capability—when presented with a problem situation in a specific mathematical domain—of communicating, reasoning, modeling, solving, and verifying propositions. Also, the index or scale developed to measure performance should reflect the student's level of maturity in that domain.

3

Implications of the NCTM *Standards* for Mathematics Assessment

Norman Webb and Thomas A. Romberg

The primary purpose of this paper is to provide an overview of NCTM's 1989 *Curriculum and Evaluation Standards* and their implications for mathematics reform. This is followed by a discussion of (1) some underlying assumptions about what it means to know mathematics and (2) ways of organizing mathematical knowledge into conceptual fields. Then, criteria for assessment, compatible with and supportive of the curriculum standards, are presented. Finally, three examples of alternative assessment techniques are given that correspond to the intent of the evaluation standards and illustrate forms of assessment that are applicable in evaluating the curriculum standards.

The National Council of Teachers of Mathematics (NCTM) Commission on Standards for School Mathematics was created in 1986 and charged with the development of a set of curriculum standards concerning (1) the mathematics that ought to be incorporated into quality school mathematics programs and (2) the instructional conditions necessary for students to learn mathematics. In addition, the Commission was asked to develop standards for both the evaluation of a school program based on the new curriculum standards and on student performance in light of those curriculum standards. During the sum-

mer of 1987, four working groups met for a month and drafted
four sets of standards: one each for grade levels K–4, 5–8, and
9–12, and one on evaluation. Drafts of the standards were
distributed to NCTM members during the 1987–88 academic
year for their review and commentary. During the summer of
1988, the working groups met again to finalize the standards
based on the feedback received and to produce a final document that was officially presented for implementation at the
1989 NCTM Annual Meeting in Orlando, Florida.

The *Curriculum and Evaluation Standards for School Mathematics* (NCTM, 1989) provides a new vision for the K–12 curriculum. That vision prescribes that mathematical knowledge,
because of its dynamic and multiplex nature, be acquired
through investigating, exploring, reasoning, making connections,
and communicating. The curriculum goal is for students to
know mathematics as an integrated whole, including a range of
topics many of which are interrelated by common symbols,
concepts, rules, and procedures.

In support of this vision, assessment as a means of observing what students know of mathematics needs to be seen differently from traditional forms of testing used in measuring
outcomes of present school curricula. Most multiple-choice or
fixed-choice tests, in which total scores are based on aggregating results from a set of items scored as correct or incorrect,
are designed to measure independent partitioning of mathematics rather than knowledge and the interrelationships among
mathematical ideas. The organization of these tests is based on
instructional objectives or competencies that reflect a view of
mathematics as a large collection of separate skills and concepts. In the new evaluation standards, assessment is viewed
as integral to instruction, with the primary role of improving
student learning. In this way assessment becomes a process of
understanding the meaning students give to mathematics—its
concepts, its procedures, and the ways problems are solved,
the reasonings used, the means of communication, as well as
how one comes to appreciate the mathematical enterprise.

Like mathematical knowledge, assessment also is dynamic
and involves a variety of approaches. Assessment is a means
for determining students' understanding of mathematical processes and the interrelationships of mathematical topics; it also
can be used to determine their ability to apply mathematics in

different situations. For the vision of the NCTM *Standards* to be realized, the new vision for assessment is necessary.

The NCTM Curriculum Standards

Mathematics is changing, and what people need to know about mathematics to be productive citizens is changing. Important factors implicated in these changes are the advances in technology, such as the prevalence of computers and calculators, and the expanding use of quantitative methods in almost all intellectual disciplines. In defining what mathematics is needed, five goals are identified by the *Standards* for the K–12 curriculum. Students are to develop their mathematical power and become mathematically literate by:

1) learning to value mathematics;
2) becoming confident in one's own ability;
3) becoming a mathematical problem solver;
4) learning to communicate mathematically; and
5) learning to reason mathematically.

There are four common standards, based on these goals, that are part of each set of Standards for each grade grouping: mathematics as problem solving, mathematics as communication, mathematics as reasoning, and mathematical connections. Positioning these standards as the first four of each set attests to their importance and their relevance to all instruction. Although not stated directly as standards, the valuing of mathematics and confidence in doing mathematics are emphasized throughout the descriptions of the standards and the suggested approaches to teaching. Focusing on problem solving, communicating, reasoning, and connections as standards for all three grade groupings recognizes that these will be attained over a period of years as a result of their reinforcement both within and across grade levels.

Solving problems, communicating, and reasoning via mathematics are not independent of each other but develop concurrently through the interaction of each with the other. The development of these mathematical abilities should be viewed as degrees of maturation within each process. Students come to kindergarten already possessing problem-solving strategies for finding answers about situations, words for describing situations, and forms of thinking about situations. Over the school years, through additional experiences, these strategies can be further developed, new strategies learned, and more sophisticated problems solved. The intent of the NCTM *Standards* is for the mathematics curriculum to become the means for expanding students' existing knowledge for introducing students to additional forms of mathematical thought; and for developing their power to use mathematics as a means of abstracting the world, interpreting the world, working within the world, and increasing their knowledge of the world.

The approach taken and the topics covered within each grade category of the *Standards* varies and is affected by the developmental level of students and the inherent structure of mathematics. In Grades K–4, the authors of the *Standards* argue that the empirical language of the mathematics of whole numbers, common fractions, and descriptive geometry, derived from the child's environment, should be emphasized. In these grades, a foundation for all further study of mathematics is firmly established. Mastery of computational algorithms has generally been considered a primary objective in the current curriculum for the lower grades. Skill and proficiency in calculating by using paper-and-pencil algorithms are important indicators of success in the curriculum. The *Standards*, on the other hand, maintains that the use of paper-and-pencil algorithms is only one among several forms of computing. In fact, depending upon the problem situation in which a computational answer is sought, one may need to estimate an answer or find an exact answer. If the latter, then one again has choices, depending on the context. One choice is to calculate mentally, a second is to use a paper-and-pencil algorithm, and another is to use a calculator. Thus, students need to learn *all* computational procedures—estimation, mental arithmetic, paper-and-pencil algorithms, and calculator uses. It is *as* important to be able to *choose among different means of computation* as it is to achieve appropriate answers.

Along with using number to describe the world empirically, it is necessary in the lower grades to develop a sense of space and knowledge of the basic concepts and rules for building geometries. Also, the underpinnings of the descriptive and inferential sciences of statistics and function that will be developed in later grades need to be introduced and experienced in Grades K–4. Throughout the process, learning mathematics should involve exploring, validating, representing, solving, constructing, discussing, using, investigating, describing, developing, and predicting.

In Grades 5–8, according to the *Standards*, the empirical study of mathematics should be extended to include other numbers beyond whole numbers, and emphasis should gradually shift to developing the abstract language of mathematics needed for algebra and more formal mathematics. The middle grades are not viewed as the culmination of the arithmetic curriculum, but are seen as a transition leading to more advanced mathematics. In this sense, the number of topics covered by all students should be increased to include significant work in geometry, statistics, probability, and algebra. The study of arithmetic skills should not be carried out in isolation but should be driven by subject matter provided in these other areas. Work in the middle grades should lead to students thinking quantitatively as well as spatially. There should be an increasing understanding of mathematical structure so that students become more aware of the relationships within and among operations, numbers, spatial figures, and other forms of representation.

In high school, Grades 9–12, students are assumed to have had the mathematical experiences of a broad, rich curriculum and to have reached some degree of computational proficiency. The emphasis of the curriculum for these later grades should be shifted from paper-and-pencil procedural skills to conceptual understanding, multiple representations and connections, mathematical modeling, and mathematical problem solving. In pursuing these, lessons should be designed around problem situations posed in an environment that encourages students to explore, to formulate and test conjectures, to prove generalizations, and to communicate and apply the results of their investigations. An important goal for the high school grades is for students to become increasingly self-directed learners, through experience in instructional programs designed to foster intellectual curiosity and independence. Although there should be variation in the depth and breadth of coverage, all students taking at least three years of high school mathematics should be exposed to algebra, functions, geometry, trigonometry, statistics, probability, discrete mathematics, the conceptual underpinnings of calculus, and mathematical structure.

Throughout the curriculum, as these topics are integrated across courses, students should become aware of the structure of mathematics and be able to recognize and make the connections among topics. These connections include forming mathematical representations of problem situations and the ability to distinguish among equivalent forms of representations.

Evaluation Standards and Assessing Mathematics

The fourteen NCTM Evaluation Standards (see Appendix A) are divided into three groups. In one group are the seven student assessment standards that describe what is to be observed and measured in the process of understanding what mathematics students know. These state that in order to adequately test mathematical knowledge, assessment needs to measure knowledge of mathematics as an integrated whole, conceptual understanding, procedural knowledge, problem solving, reasoning, communication, and mathematical disposition. A second group comprises three general assessment standards that present principles for judging assessment instruments. Inherent in the general assessment standards is an assumption that all evaluation processes should use multiple assessment techniques that are aligned with the curriculum and consider the purpose for assessment. A third group comprises the four standards that identify what should be included in evaluating a mathematics program. One purpose for program evaluation is to obtain relevant and useful information for making decisions about curriculum and instruction. These four evaluation standards provide indicators of a mathematics program consistent with the Curriculum Standards, the focus for examining the instructional resources of a mathematics curriculum, the focus for examining instruction and its environment to determine a mathematics program's consistency with the Curriculum Standards, and provisions for program evaluation, to be planned and conducted by an evaluation team.

For the purpose of reflecting on the implications of the NCTM Evaluation Standards for assessing mathematics, this paper will focus on the final group, the three standards that address general assessment criteria: alignment, multiple sources of information, and appropriateness. It is these three standards that can be used to justify the consideration of new or alternative forms for assessing mathematics other than just changing the content of tests.

Alignment of Assessment

In order for methods of assessment to be aligned with the NCTM Curriculum Standards, their spirit and their goals, the assessment methods need to conform to the *Standards* at the instructional level, program level, and mathematical domain level.

Instructional Alignment. The assessment method will be aligned with instruction to the extent that it covers the breadth of topics taught and provides information about the full range of student outcomes expected. The concept of alignment is used instead of that of validity to stress the dynamic nature of assessment and the need to use multiple sources for information. Where traditionally the validity of one instrument or test is analyzed, the NCTM Evaluation Standards indicate that any one instrument will be insufficient to measure the full intent of instruction based on the NCTM *Standards*. Thus, in bringing the assessment methods into agreement or into alignment with instruction, it is necessary to consider the variety and range of assessment methods being used.

It is also important to consider the learning environment and its expectations for the use of technology. Alignment means that if certain materials or equipment are being used in instruction and are a part of the mathematical experiences of the learners, then these materials and equipment should be used in assessment. For example, the Curriculum Standards note that calculators are one of several means for computing. Calculators also are a means of exploring and investigating mathematical ideas. Thus, for assessment to be aligned with instruction represented in the *Standards*, learners should at some time during assessment have the option of using calculators to do computations and to investigate other mathematical ideas.

Program Alignment. In addition to instructional alignment, assessment methods should be aligned with the total K–12 mathematics program and conform to the expectations and goals that students are to have attained at the completion of eleven or twelve years of mathematics. This is referred to as program alignment. The NCTM *Curriculum Standards* describe what it is that students should know about mathematics, about mathematical concepts and procedures, and about applying reasoning and communication for the purpose of understanding and solving problem situations. The goals for a K–12 program, as described in the *Standards*, cannot be achieved by aggregating learning that has been partitioned into eleven or twelve segments, but will be achieved only through having a common thrust across grades with each building upon the knowledge developed in previous grades and with each leading toward constructing a complete knowledge of mathematics. There must be an articulated program so that students progress through the grades with their knowledge of mathematics maturing in a logical and deliberate sequence.

When an assessment method is in alignment with the teaching program, the method will measure what students know about mathematics that assures the desired level of knowledge in order for them to be productive citizens throughout their lifetimes. For example, administering timed tests of basic addition and subtraction facts would be part of a program only if this method is supported by collecting evidence on how well students can provide an explanation of their efficient thinking for determining answers in a number of ways. If timed tests are used only to assess the students memorization of facts without understanding, then timed tests, as a measure of learning, are not in alignment with the program goals that project for students the development of a number sense and the foundation for developing knowledge of the real number system.

The set of assessment methods used at any specific grade level or within any specific course may lack program alignment simply due to the omission of methods of gathering evidence on an aspect rather than because a method is being used that does not coincide with program goals. A major goal of program compliance with the NCTM *Curriculum Standards* is for students to achieve the ability to communicate mathematically. This means that students need to be able to use the language of mathematics to talk, to write, to listen, and to read mathematics. Students are to be engaged in the communication of mathematics at all grade levels. Student assessment within each grade should include some procedures for observing and measuring the development of student ability to communicate. The assessment situation should match as closely as possible the desired outcomes and normal progression toward the program goal. The assessment of students' communication skills can be made via interviews, or by some other means of having students explain their thinking.

Mathematical Domain Alignment. The third type of alignment for assessment methods is alignment with the field of mathematics, its structure, and organization. The goal is for students to construct a body of knowledge that will result in their having the power of mathematics. If students are to acquire the knowledge of mathematics described in the NCTM *Standards*, they need to know the concepts, symbols, procedures, reasoning, and processes of mathematics as well as their structure and their interrelationship. Students also need to be able to apply mathematics to situations that add meaning to these symbols and concepts.

For an assessment method to be aligned with the field of mathematics means that students are tested on mathematics in a way that is compatible with the structure of mathematics and how mathematics exists within the minds of students. Consideration of the structure of mathematics in constructing assessment methods affects how tasks are designed and chosen, how tasks are administered, the desired form of response, what rules are followed to make judgments about responses, and how information is aggregated and reported.

One example of a strategy for constructing assessment instruments that is aligned with a conception of the field of mathematics is the *domain knowledge* strategy (Romberg, 1987). The domain knowledge strategy is based on Gerard Vergnaud's (1982) notions about *conceptual fields*. His theories are based on the philosophic premise that the power of mathematics lies in the fact that a small number of symbols and symbolic statements can be used to represent a vast array of different problem situations. If a set of symbols represents a closely related set of concepts, referred to as a "conceptual field," then this monitoring framework should allow one to determine the degree of knowledge a student or group has acquired with respect to that domain.

The properties of a conceptual field are:

- a set of situations that makes the concept meaningful;
- a set of invariants that constitutes the concept; and
- a set of symbolic situations used to represent the concept, its properties, and the situations it refers to.

For example, the related mathematical concepts of addition and subtraction of whole numbers has been defined by Vergnaud as the conceptual field *additive structure*. Such fields are derived in four steps.

1. The symbolic statements (e.g., $a + b = c$ and $a - b = c$; where a, b, and c are natural numbers) which characterize the domain are identified.

2. The implied task (or tasks) to be carried out are specified. For addition and subtraction, this involves describing the situations where two of the three numbers, a, b, and c, in the statements are known and the other is unknown.

3. Rules (invariants) are identified that can be followed to represent, transform, and carry out procedures to complete the task (e.g., find the unknown number using one or more of such procedures as: counting strategies, basic facts, symbolic transformations such as $a + [\ ] = c$ which implies $c - a = [\ ]$, computational algorithms for larger numbers, and others).

(Note that in the first three steps only formal aspects of a mathematical system are considered.)

4. A set of situations are identified that have been used to make the concepts, the relationships between concepts, and the rules meaningful (e.g., join-separate, part-part-whole, compare, equalize, fair trading).

The result of these four steps yields a map (a tightly connected network) of the domain knowledge. It is this map that is used as a framework for assessment, instead of other possible frameworks, such as a content-by-behavior framework. In addition to the additive structure, other conceptual fields would include multiplicative structure, proportional reasoning structure, probability structure, spatial structure, logical structure, relational structure, iterative (discrete) structure, measurement structure, algebraic structure, integral structure, and the differential structure. These structures overlap in that some will use common symbols, concepts, and rules. Problem situations that will be used also apply across the different fields.

One of the implications of using a domain knowledge strategy is that the set of situations that gives meaning to the concepts and rules is equally as important as learning to follow the procedural rules. Knowledge of a domain is viewed as forming a network of multiple possible paths and not partitioned into discrete segments as implied by such models as the content-by-behavior model. Over time, the maturation of a person's functioning within a conceptual field should be observed by noting the formation of new linkages, the variation in the situations the person is able to work with, the degree of abstraction that is applied, and the level of reasoning applied. What is important for alignment is that the assessment methods conform to some conception of the field being assessed.

Mathematics is plural and represents a field of study composed of several domains. To increase the knowledge of mathematics can be interpreted as maturing in the knowledge of each domain. However, there are common ideas, symbols, and procedures that interconnect the different domains. One challenge for better understanding the interrelationship of domains is to construct a map of how specific domains are related. Such a map would be very valuable for guiding both interaction and assessment.

Multiple Sources of Information

In the use of multiple forms of assessment, inferences about what a student knows must be based on confidence in how the evidence from different sources converges to support a single conclusion. Traditional notions of reliability are not as meaningful or as applicable when trying to determine what someone knows and when making instructional decisions. Any one source of measurement, such as a test, will naturally have built-in error as a measure of what mathematics a person knows, simply because mathematical knowledge is multifaceted. It is also not feasible to expect teachers and others who develop assessment instruments to have the time and resources to develop highly reliable tests in the classical sense. For instructional purposes, assessment should be viewed as an ongoing process of gathering information for making instructional decisions and for reporting the outcomes of instruction in relation to the domain of mathematical knowledge. Confidence in one source of evidence can only be achieved by supporting evidence from other sources. To assess the intent of the Curriculum Standards, many different forms of assessment will have to be used.

Appropriate Assessment Methods and Uses

A general assessment criterion is that the assessment method be appropriate to the use that will be made of the results. As with any assessment, the method needs to coincide with the purpose for doing the assessment and with the developmental and mathematical maturity of the students. There are many different purposes for assessment such as grade-to-grade promotion, graduation requirement, diagnosis, instructional grouping, and program evaluation. Assessment for the purpose of judging the strength and weaknesses of a school district mathematics program will need to use methods different from those used for assessment for the purpose of assigning grades to individual students. In deciding on any method of assessment, the purpose of the assessment needs to be an important consideration.

The more closely the form of assessment matches the level of mathematical maturation of the learner, the more useful will be the information obtained. A key principle of assessment is to determine what mathematics the learner knows. This is done by locating where the student is on the map of mathematical knowledge by noting what meaning the learner gives to concepts and symbols, what procedures the learner knows and can use, and how the learner is able to reason, solve problems, and communicate. In order to locate the individual precisely, the assessment instrument must be sensitive to the distinctions that the learner makes. This requires refined assessment instruments. It implies that manipulatives should be a part of the assessment environment when assessing the mathematical knowledge of primary age learners. Instruments that distinguish among forms of abstract knowledge should be useful in assessing the knowledge of eleventh and twelfth graders who have experienced a curriculum corresponding to the *Standards*.

Criteria for Judging Assessment Instruments

In judging assessment instruments for meeting the main criteria listed in the Evaluation Standards, four points need to be considered.

1.   The assessment instrument should provide information that will contribute to decisions for the improvement of instruction.
2.   The assessment instrument should be aligned with the instructional goals, the goals for the overall program, and a holistic conceptualization of mathematical knowledge.
3.   The assessment instrument should provide information on what a student knows.
4.   The results from one assessment instrument should be such that when combined with results from other forms of assessment, a global description is obtained of what mathematics a person or group knows.

Three Examples of Alternative Assessment Tasks

The components most commonly encountered in large scale assessments are fixed-choice items, multiple-choice items, and items requiring short answers. Classroom assessment is thought of as giving students a set of problems in symbolic form—such as an equation, sequence of numbers, or words—with each problem requiring a number as an answer. Other than these, there are few alternative forms of assessment of mathematics currently in use. There are few that provide information on the communication of mathematics, on the understanding of mathematics as an interrelated set of ideas, and on how well the learner is able to gain meaning from the situation.

Recently certain interesting methods of assessment have been explored that show some promise and conform to the vision of the *Standards*. Some of these newer methods have been developed in other countries as problems of reform in those countries are being addressed. Examples from three different sources are given below. These examples have been selected to illustrate different means of assessment that seem to adhere to the spirit and recommendations in the NCTM *Standards*. The three examples were also selected to reflect what can be done at different grade levels. However, each assessment method discussed should be applicable to a broader grade range than the one presented.

Grades K–4 Assessment

The work of Carpenter, Fennema, Peterson, and Carey (1987) on the Cognitively Guided Instruction Project illustrates making assessment a part of instruction. A guiding principle for Cognitively Guided Instruction is that instructional decisions should be based on careful analyses of students' knowledge and the goals of instruction (Carpenter & Fennema, 1988). Instruction should be appropriate for each child's level of knowledge and facilitate growth on successive levels. This requires that individual students be assessed at regular intervals and that instruction planning be based on the results of assessment. In Cognitively Guided Instruction, it is important to assess not only whether a learner can solve a particular problem, but also how the learner solves the problem. This is very compatible with the emphasis of the NCTM *Standards* on problem solving and reasoning.

Three methods of assessment are being explored by the Cognitively Guided Instruction Project (Carpenter, Fennema, & Peterson, unpublished): assessing children in individual interviews, assessing children during group instruction, and spot-checking assessment during seatwork. Clearly, interviewing or observing each student in a class requires some form of organization for assessment so that all students are observed or interviewed during the assessment period. There also needs to be some means for systematically recording learners' responses. It is important when using these methods for the teacher to have knowledge of the classification of problems and of children's strategies. In this way, the teacher is able to ask questions or to structure the situation to more appropriately conform to the learner's development and mathematical maturity. In viewing assessment in this way, it is not necessary to use large batteries of problems in order to make important and relevant decisions for instruction. What is important is to give the learner a problem that matches or nearly matches his/her level of knowledge. In interviews, there is more flexibility to make adjustments, such as leading the student in a given direction.

The following protocol was presented as a way of indicating whether a child is ready to proceed from Counting All (given two sets of objects, the student counts all objects, beginning with one, to determine the sum) to Counting On (given two sets of objects, the student counts beginning with the number in one set).

> TEACHER: "There are 6 pennies in the bank." (The teacher places the pennies in the bank without counting each one.) "How many pennies will be in the bank if we put in 2 more?"

Paul begins to count on his fingers, "1, 2, 3, 4, 5, 6" to establish the number 6, then hesitates and counts 2 more fingers. He looks at the 8 fingers and says "Eight."

> TEACHER: "Do you think you could solve the problem without counting on all 8 fingers?"

> PAUL: No response.

> TEACHER: "When you count, what number comes after 6?"

> PAUL: "Seven comes after 6."

> TEACHER: "Right. Suppose we had 7 pennies in the bank and we add 1 more penny. How many pennies would we have? Can you think of the number that is 1 more than 7 when you count?"

> PAUL: "Well, . . . 7, 8. Eight comes after 7."

> TEACHER: "Good. Let's put 7 pennies in the bank." (Teacher places chips in groups.) "If we put 2 more pennies in the bank, can we figure out how many pennies there will be altogether?"

> PAUL: "Seven, (pause) 8, 9. There are 9 pennies."

To assess whether Paul could independently Count On, a similar problem using the number fact 5 + 2 was given. (Carpenter, Fennema, & Peterson, unpublished, pp. 7–8)

The process described in this protocol is an example of a teacher leading a student through a situation. Through the process of trying to determine whether Paul could approach the problem by some means other than concretely using fingers, the teacher, by asking questions, is able to determine if the student has some of the knowledge necessary to proceed. In this situation, a student could make a simple counting error and produce a wrong answer. Without interviewing, a teacher may reach the wrong conclusion about the student's readiness to use the Count On strategy.

Grades 5–8 Assessment

An example of using a situation to partially assess middle school students' knowledge of mathematics as communication and reasoning and knowledge of number comes from the Shell Centre for Mathematical Education (Swan, 1985) in England. Figure 3–1 illustrates a specimen examination question from a module aimed at developing the performance of children in interpreting and using information presented in a variety of familiar mathematical and nonmathematical forms. Figure 3–2 gives the marking scheme for the question.

This particular example requires the learner to describe in words the relationship of one form of representation (a map) to another form of representation (a graph). Then in sketching a graph, the learner has the opportunity of modeling a situation graphically. In constructing the graph, some proportional reasoning is required. Of note is that more than one question is derived from a situation. This provides an opportunity to observe the interrelatedness of different forms of representation. The learner will need to have a knowledge of concepts in order to describe the car journey. Procedural knowledge is required in reading the map and graph in order to get information from each. In order to sketch the graph, reasoning is required to determine the speed of the car in relation to the time. The marking scheme then gives some indication of what a learner knows by indicating a global score based on how well that student is able to bring everything together.

This example demonstrates a form of assessment aligned with a conception of mathematics that involves different forms of representations within the conceptual field of proportional reasoning. A learner who is able to score high on this situation shows a good understanding of speed as a form of proportion by being able to derive meaning from the map and graph. By providing a score for the different parts of the situation, it is

THE JOURNEY

The map and the graph below describe a car journey from **Nottingham** to Crawley using the M1 and M23 motorways.

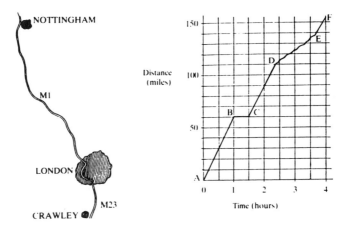

(i) Describe each stage of the journey. making use of the graph *and* the map. In particular describe and explain what is happening from A to B; B to C; C to D; D to E and E to F.

(ii) Using the information given above. sketch a graph to show how the **speed** of the car varies during the journey.

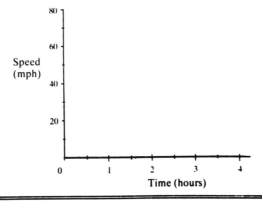

Figure 3–1. One examination question from the Shell Centre for Mathematical Education. (Swan, 1985, p. 13). Reprinted with permission.

THE JOURNEY . . . MARKING SCHEME

(i)   **Interpreting mathematical representations using words and combining infor-
      mation to draw inferences.**

Journey from A to B	'Travelling on the M1'	1 mark
	'Travelling at 60 mph' (± 5 mph) or 'travels 60 miles in one hour'	1 mark
Journey from B to C	'Stops' or 'At a service station' or 'In a traffic jam' or equivalent	1 mark
Journey from C to D	'Travelling on the motorway'	1 mark
	'Travelling at the same speed as before' or 'Travelling at 60 mph (± 5 mph)' or 'Travels 50 miles in 50 minutes' (± 5 mins.)	1 mark
Journey from D to E	'Travelling through London'	1 mark
	'Speed fluctuates', or equivalent. eg: 'there are lots of traffic lights'. Do not accept 'car slows down'.	1 mark
Journey from E to F	'Travelling on the motorway' or 'Travelling from London to Crawley'.	1 mark

(ii) **Translating into and between mathematical representations.**

For the general shape of the graph, award:

1 mark   if the first section of the graph shows a speed of 60 mph (± 10 mph) reducing
           to 0 mph.

1 mark   if the final section of the graph shows that the speed increases to 60 mph (±
           10 mph) then decreases to 20 mph (± 10 mph) and then increases again.

For more detailed aspects, award:

1 mark   if the speed for section AB is shown as 60 mph and the speed for section CD
           is shown as 60 mph (± 5 mph).

1 mark   if the changes in speed at 1 hour and 1½ hours are represented by (near)
           vertical lines.

1 mark   if the stop is correctly represented from 1 hour to 1½ hours.

1 mark   if the speed through London is shown as anything from 20 mph to 26 mph or
           is shown as fluctuating.

1 mark   if the graph is correct in all other respects.

A total of 15 marks are available for this question.

Figure 3–2.   Scoring scheme for examination question given in Figure 3–1. (Swan,
                1985, p. 13). Reprinted with permission.

easier to determine what a student knows and what links have formed. Having students write a response to the problem-solving process provides an occasion to observe the use of language to describe a situation given in pictorial form using a map and a graph.

Grades 7–12 Assessment

In The Netherlands, mathematics education is in a state of transition just as it is in this country. Out of the reform effort has come a course referred to as Mathematics A (de Lange, 1987) that is aimed at students who are expected to pursue studies at the university in disciplines where mathematics is needed only as a tool. Many of the students who take Mathematics A will specialize in economics, social sciences, and medicine. One form of assessment used is written timed tests. These are classified as follows:

1. Class 1. Exercises without context, or with hardly any context
2. Class 2. Exercises with a substantial use of context
    a. Exercises strongly resembling the exercises from the booklet[1]
    b. Exercises resembling those of the booklet somewhat, but not strikingly
    c. Exercises not resembling those of the booklet.

One sample form of assessment included in the Mathematics A materials is a "Two-Stage" task. In this learners are given a situation and asked to respond to as many of the questions as possible in a traditional timed written test. The first half of the test consists mainly of open-ended questions. The second half of the test may include essay questions. The results are scored and then returned to the student.

For the second stage, students are provided information on their scores and on the gross mistakes they made in the first stage. Then the students are asked to repeat their work on the

[1]A booklet is a separately bound unit based on a realistic set of problems. The course is organized around several booklets, each of which takes two to four weeks to complete. (de Lange, 1987). Reprinted with permission.

problem situation at home where they have no restrictions and are completely free to answer the questions as they choose. Students may be given as much as three weeks to do this. Then students are scored on both stages.

An example of a problem given is the Forester Problem. This is to test learners' knowledge of matrices. (de Lange, 1987, pp. 187–89)

FORESTER PROBLEM

A forester has a piece of land with 3,000 Christmas trees. Just before Christmas he cuts a number of trees to sell them. The forester distinguishes three classes of length: S, M, L trees. The small trees have just been planted and have no economic value, the medium trees are sold fl. 10,– a piece, and the large ones for fl. 25,–. He has, just after Christmas, 1,000 small, 1,000 medium, and 1,000 large trees. All these grow uneventfully until just before next Christmas. From experiences of colleagues, he knows approximately about the growth per year:

40% of the small trees become medium
20% of the medium trees become large
Or, in a GRAPH:

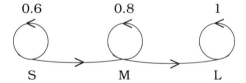

This graph may be represented by a GROWTH-MATRIX G:

$$
\begin{array}{c}
\text{from} \\
\begin{array}{ccc}
S & M & L
\end{array} \\
\text{to} \left\{ \begin{array}{ccc}
.. & .. & .. \\
.. & .. & .. \\
.. & .. & ..
\end{array} \right\} \begin{array}{c} S \\ M \\ L \end{array}
\end{array}
$$

1. Complete the matrix G.
2. Calculate the composition of the forest just before the next Christmas (using G).
3. After cutting medium and large trees and planting

small trees, the forester wants his starting population B (1,000 of each) back. How many of each kind should be cut and planted?

4. Cutting one tree costs fl. 1,–; planting one tree costs fl. 2,–; what will be the forester's profit this Christmas?

The forester wonders whether the above strategy is the most profitable one. He considers two other strategies, so he has the choice of the following three strategies:

I.   Cut after one year and plant so as to get your starting population back.
II. Cut after two years and plant back so as to get the starting population.
III. Cut after one year the large trees only (leaving 1,000) and replant the same number of small trees: repeat this the second year.

5. Which of the above three strategies is the most profitable per year?

The forester considers the use of fertilizer to make the trees grow faster. There exists a fertilizer that, according to the producer, might lead to the following growth-matrix:

$$G = \begin{Bmatrix} 0.6 & 0 & 0 \\ 0.4 & 0.5 & 0 \\ 0 & 0.5 & 1 \end{Bmatrix}$$

6. Explain why the trees grow faster with this mode.

The forester likes to use the fertilizer but has some doubts whether it would not be possible to get the start population B back after each Christmas, because getting back the B-population is essential to him.

7. Will it be possible to get back the start population B when the matrix G is of issue (after one year)?
8. The forester decides to thin the fertilizer such that to get his B-population back every year he only has to cut large trees and to replant the same

number of small trees. Can you suggest any Matrix G that will have the desired effect?

9.   Using all information available, what would you
     advise the forester?

Another forester prefers five length-classes:

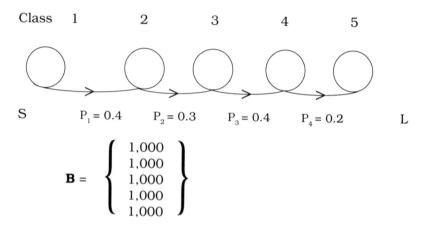

Class   1   2   3   4   5

S $P_1 = 0.4$ $P_2 = 0.3$ $P_3 = 0.4$ $P_4 = 0.2$ L

$$\mathbf{B} = \left\{ \begin{array}{c} 1,000 \\ 1,000 \\ 1,000 \\ 1,000 \\ 1,000 \end{array} \right\}$$

10. Write an essay including:
     — the growth-matrix in this situation
     — the effect on the total population after one year
     — the possibility to get B back
     — if this is impossible, change one of the entries of
       the matrix in order to make it possible
     — the effect if a tree manages to grow in one year
       from class 3 to class 5.

11. Find the matrix for the general case:

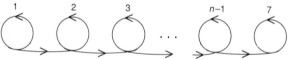

How can you conclude from the matrix whether
getting B is possible or not?

12. What are the limitations of the model? What re-
     finements would you like to suggest?

13. (See question 5) The third strategy was: Cut after

one year the large trees only (leaving) and replant the same number of small trees; the same is done the next year(s). What will be the effect of this strategy in the long run?

The Forester Problem illustrates the development of several questions from one situation. The questions are sequenced by difficulty and complexity. The final questions require the student to make generalizations and draw conclusions from the previous work. One difficulty in giving such an assessment task is the scoring. De Lange noted that teachers approach the scoring in different ways. One approach was to read the complete work, mark positive and negative points, and then give a grade. It is possible that such a situation could provide some rich data on motivated students who complete the task.

The three examples given above were presented to generate thought and dialogue regarding assessment. Each demonstrates a means for assessing what students know about an area of mathematics. The Carpenter, Fennema, and Peterson protocol for a teacher-student interview illustrates a teacher using information from the student and leading the student to a more advanced counting strategy. This requires a deep understanding by the teacher of the procedures and of the student's thinking process. But in the interview situation it becomes apparent that the student begins to make links between existing knowledge and computation of the sum by using Counting On. The use of this type of approach by the teacher supports the K–4 Curriculum Standards by teaching students different strategies through using known procedures. The teacher interview is a viable assessment approach for the Evaluation Standards, because the teacher obtains information on the students' procedural knowledge that can lead to further development of the new strategy.

The Swan examination question illustrates the assessment of students' knowledge of different forms of representation. This approach allows the teacher to understand better what the student is able to do with proportional reasoning. Having the student respond verbally and graphically provides an opportunity to observe the link between the concepts of speed and the different forms of the language of mathematics. Awarding scores on parts of the task indicates knowledge of the same concept in different contexts created at different stages of the journey. The de Lange Forester Problem illustrates an assessment situation that extends to more than one approach—a timed test and an extended test. For one situation, multiple questions are asked. These questions cover a span of knowledge on the use of matrices, from applying basic operations to being able to generalize to n dimensions. Having students respond to a number of questions regarding a given situation in different ways under different conditions provides the opportunity to observe the use of multiple sources of information and to ensure a clearer understanding of the depth of knowledge the student has.

The three examples discussed in this paper generate more questions than answers. The examples correspond to the spirit of the NCTM *Standards* and provide evidence of the existence of some alternative forms of assessment. But the challenge exists for those in testing and evaluation to make use of such examples and develop others that are both aligned with curriculum and provide useful and accurate information. In doing this, the forms of assessment must be compatible with the structure and nature of mathematics and aligned with instruction and the curriculum program. For teachers, the challenge is (1) to become comfortable with the different means of assessment and (2) to apply these tools for building instructional strategies and an understanding of what students know.

4

Curriculum and Test Alignment

Thomas A. Romberg, Linda Wilson,
'Mamphono Khaketla, and Silvia Chavarria

The purpose of this chapter is to report on information gathered from two studies related to the reality of Evaluation Standard 1: Alignment of the NCTM *Standards* (see Appendix A). In the first study, the six standardized tests most widely used at state and district levels in schools in the United States are examined to determine whether or not they are appropriate instruments for assessing the content, process, and levels of thinking called for in the *Standards*. The results show that the tests are not appropriate. They are found to be generally weak in five of the six content areas and in five of the six process areas. Furthermore, the tests place too much emphasis on procedures and not enough on concepts. In the second study, we conduct an examination of items and tests from newly developed state tests and foreign tests. It is clear that there are test items currently in use and some being developed that provide the kind of breadth of content and depth of knowledge cited in the *Standards*.

PART 1: THE STUDY OF SIX STANDARD TESTS

Background

Romberg, Wilson, and Khaketla's 1989 study "An Examination of Six Standard Mathematics Tests for Grade 8" followed an earlier large-scale questionnaire survey conducted by Romberg, Zarinnia, and Williams (1989). The survey was conducted to find out from Grade 8 teachers how mandated testing influenced their teaching of mathematics. It sought information from teachers about district and state tests that their students took, the amount of time they spent on testing and preparing students to take tests, how they used test results, what their views were about the effects of the tests, how the tests influenced their teaching, and what they perceived as the influence testing had on the mathematics curriculum.

The results of the study indicate that nearly 70 percent of the teachers report that their students take a mandated test at either the district or state level or both. Secondly, because teachers know the form and character of the tests their students take, most teachers make changes in their teaching to reflect this knowledge. Thirdly, the changes teachers make in their classroom practice tend to contrast with the recommendations made by NCTM's *Curriculum and Evaluation Standards for School Mathematics* (1989). For example, the *Standards* recommended more activities involving the use of calculators in the classroom (p. 8; p. 75). However, about 25 percent of the teachers report that they decreased emphasis on calculator activities because students cannot use calculators on standardized tests; less than 10 percent reported an increased use of calculators in their classrooms.

One of the survey questions asked teachers to list the tests that their schools used. Six commercially developed tests were listed as the most widely used in Grade 8, both at the district and state level: The California Achievement Test (CAT) (1985), The Metropolitan Achievement Test (MAT) (1986), The Stanford Achievement Test (SAT) (1982), The Science Research Associates Survey of Basic Skills (SRA) (1985), The Comprehensive Test of Basic Skills (CTBS) (1989), and The Iowa Test of Basic Skills (ITBS) (1986). The analysis of these six tests serves as the basis for the findings in the report of the first study.

Purpose

The six tests were analyzed to (1) determine whether they reflect the recommendations made in the *Standards*, and, if not, to (2) make recommendations to the test developers for revising their tests. The rationale for this study is based on the fact that since a number of teachers reported changing their teaching to reflect their knowledge of what is tested, the best way to ensure a shift in emphasis in teaching is to change the test. It is a rationale based directly on the statement, "To change the curriculum, change the test," for which considerable evidence exists (e.g., Edelman, 1980; Millman, Bishop, & Ebel, 1965; Romberg, Zarinnia, & Williams, 1989; Stanley & Hopkins, 1981).

A major argument against standardized tests has been their failure to assess higher-order skills; rather, such tests emphasize computations, recognition, and other lower-order thinking skills (Meier, 1989; Putnam, Lampert, & Peterson, 1989). This latter type of assessment is in contrast to the major theme of the *Standards*—Problem Solving—which is an all-encompassing higher-order skill. However, before urging test developers to change their tests, it is necessary to determine whether current tests evaluate the achievement of the objectives set out in the *Standards*. The intent of this study of six standard tests, therefore, is to make that determination.

Design

The mandated tests under study were Grade 8 tests. Therefore, the Standards for Grades 5–8 formed the basis for the classification of items.

Each item on each test was classified within three areas: (1) the *content* it tests; (2) the *process* required to respond to the item; and (3) the *level* of the response required.

The item was first categorized into one of the following seven *content* areas described in the *Standards* (the numbers in parentheses refer to their position in the Grade 5–8 Standards):

Number and Number Relations (5)

Number and Number Theory (6)

Algebra (9)

Statistics (10)

Probability (11)

Geometry (12)

Measurement (13)

Each item was categorized into one of six *process* areas described in the *Standards*:

Communication (2)

Computation and Estimation (7)

Connections (4)

Reasoning (3)

Problem Solving (1)

Patterns and Functions (8)

Finally, each item was classified into one of two *levels*, procedure or concept, according to whether the response to the question required procedural or conceptual knowledge.

A matrix (Appendix B) was developed and used to classify each test item according to the three classifications. For example, for the CTBS test, Item 1, "8.685 – 2.150," is classified as a *computation* problem that tests *number and number relations*: to work it requires *procedural knowledge*. Totals were generated for each test, and the results are discussed in the results section.

Two raters did the classification. A test was chosen at random and analyzed by both raters working together. Then a different test was picked and individually analyzed by each rater. Their results were compared for interrater reliability. No significant differences in the individual ratings were found, and the remaining tests were divided among the raters for individual analysis.

RESULTS OF THE ANALYSIS

Problems

Two problems were encountered in categorizing the test items. First, in many cases, an item was put into a certain category even though in fact it did not reflect the true spirit of the *Standards*. For example, an item from the ITBS has the following stem: "How would you write 6 thousandths as a decimal?" This was categorized under the process area, Communication, though it requires a much "lower" level of communication than that described in the *Standards*:

In Grades 5–8, the study of mathematics should include opportunities to communicate so that students can:

- model situations using oral, written, concrete, pictorial, graphical, and algebraic methods;
- reflect on and clarify their own thinking about mathematical ideas and situations;
- develop common understandings of mathematical ideas and situations;
- develop common understandings of mathematical ideas, including the role of definitions;
- use the skills of reading, listening, and viewing to interpret and evaluate mathematical ideas;
- discuss mathematical ideas and make conjectures and convincing arguments;
- appreciate the value of mathematical notation and its role in the development of mathematical ideas. (NCTM, 1989, p. 78)

Second, many of the tests label a section "Problem Solving," yet the problems do not resemble the types of problems described in the *Standards* as problem solving. Nearly all of the problems so labeled in the tests were routine word problems, such as: "Brett correctly answered 24 out of 25 questions on a science test. What percent of the questions did Brett answer correctly?" (SRA, Level 36, Form P). The *Standards*, on the other hand, call for nonroutine problem situations which "are much broader in scope and substance than isolated puzzle problems" and "very different from traditional word problems, which provide contexts for using particular formulas or algorithms but do not offer opportunities for true problem solving" (NCTM, 1989, p. 76).

Individual Test Results

The percent of items classified in each category for each test is given in Appendix C.

1. *SRA Survey of Basic Skills*
   Level 36, Form P
   Science Research Associates, 1985
   Chicago, Illinois
   There were 90 items in the mathematics portion of the test. Of those, the majority (82 percent) were classified in the content area of Number and Number Relations, with 7 percent each in Number Systems and Number

Theory, 4 percent in Geometry, and none in Probability, Statistics, or Measurement. In areas of process, most (91 percent) of the items were classified as Computation/Estimation, with 5 percent in Communication, 3 percent in Problem Solving, 1 percent in Reasoning, and none in Connections or Patterns and Functions. Eighty-four percent of the items were viewed as procedural and 16 percent conceptual.

2. *The California Achievement Test*
Level 18, Form E, Tests 6 & 7
CTB/McGraw Hill, 1985
Monterey, California
There were 105 items on the mathematics portion of the test. This test had a somewhat broader distribution of content areas, with 73 percent of the items in Number and Number Relations, 6 percent each in Measurement, Algebra, Probability or Statistics, 5 percent in Number Systems and Number Theory, and 4 percent in Geometry. In the process areas, most (83 percent) of the items were Computation/Estimation, 11 percent were Communication, 6 percent were Reasoning, and none were in Problem Solving, Connections, or Patterns and Functions. Most (90 percent) of the items were procedural, with 10 percent conceptual.

3. *The Stanford Achievement Test*
Advanced Level (7th ed.)
The Psychological Corporation, 1982
San Antonio, Texas
While most (64 percent) of the 118 items on the mathematics portion of the test were classified in the content area of Number and Number Relations, a broader spread existed among the other content areas. Fifteen percent of the items were classified as Measurement, 10 percent as Algebra, and 9 percent as Probability or Statistics. However, only 2 percent of the items were in Geometry, and none were in Number Systems and Number Theory. In the process categories, 38 percent of the items were in Communication, 62 percent were in Computation/Estimation, and none were in the other categories. Nearly all (92 percent) of the items were considered procedural, with only 8 percent conceptual.

4. *The Iowa Test of Basic Skills*
   Level 14, Form 7
   Riverside Publishing Co., 1986
   Chicago, Illinois
   The mathematics portion of this test, with 193 items, was nearly twice as large as the other five tests under study. However, like the others, the majority (62 percent) of the items were classified in the content area of Number and Number Relations. Measurement had 13 percent of the items, and 11 percent of the items were in Number Systems and Number Theory. Seven percent of the items were in Algebra, 4 percent were in Geometry, and 3 percent were in Probability or Statistics. Nearly all (89 percent) of the items were in the process area of Computation/Estimation, with 9 percent in Communication, 1 percent in Patterns and Functions, 1 percent in Reasoning, and none in the other process categories. Only 4 percent of the items were considered conceptual, with 96 percent procedural.

5. *The Metropolitan Achievement Test*
   Advanced 1, Forms L & M (6th ed.)
   The Psychological Corporation, 1986
   San Antonio, Texas
   This test, with 95 mathematics items, was quite similar to the others in content. Sixty-six percent of the items were classified as Number and Number Relations, 15 percent as Measurement, 8 percent as Geometry, 6 percent as Number Systems and Number Theory, 5 percent as Probability or Statistics, and none as Algebra. Like the others, most (79 percent) of the items were classified in the process area of Computation/Estimation, with 21 percent in Communication, and none in the others. Eighty-eight percent of the items were procedural, and 12 percent conceptual.

6. *The Comprehensive Test of Basic Skills*
   Level 17/18, Form A
   CTB/McGraw Hill, 1989
   Monterey, California
   Seventy-six percent of the 94 items on the mathematics portion of the test were classified as Number and Number Relations. Eleven percent were classified in the con-

tent area of Probability or Statistics, 8 percent in Geometry, 5 percent in Measurement, and none in the other areas. Most (71 percent) of the items were Computation/Estimation, with 25 percent in Communication, 2 percent each in Reasoning and Patterns and Functions, and none in the other areas. Eighty-five percent of the items were procedural, and 15 percent were conceptual.

General Results

1. Most items (between 62 percent and 82 percent, with an average of 71 percent) were found to be in the content area of Number and Number Relations, with the rest fairly evenly distributed among the other five categories.

2. Most items (between 71 percent and 91 percent, with an average of 79 percent) were found to be in the process area of Computation/Estimation, with 20 percent in Communication, and very few in the other four categories.

3. An average of 89 percent (with a range of from 84 percent to 96 percent) of the items were classified as procedural, rather than conceptual.

4. Little variation was found among the tests in terms of categorizing the items. The greatest range was found in the category of Communication, with the SRA having 5 percent of its items in Communication and the SAT having 38 percent.

5. In the category of Computation/Estimation, the majority of items were Computation, with no more than 10 percent, and in some cases 0 percent, being Estimation.

Conclusions

Romberg, Wilson, and Khaketla's 1989 study of the six standardized tests used most widely at state and district levels in schools in the United States made the following findings:

1. The items in the tests examined do not adequately cover the range of content described in the *Standards*. The great majority of items were found to be computations on numbers, or were based on

algorithmic procedures. The *Standards*, on the other hand, call for *a decreased emphasis on performing routine computations by hand*, and *increased emphasis in other areas such as problem solving, reasoning, connections, and communication.*

2. The tests do not address one of the primary content areas of the *Standards*, that is, problem solving. The *Standards* "strongly endorse" the first recommendation of *An Agenda for Action*, which states that problem solving must be the focus of school mathematics (NCTM, 1980, p. 6). Further, the *Standards* consider problem solving to involve much more than routine word problems. According to that definition, an average of only 1 percent of the items on the tests were categorized as problem solving, with a range of from 0 percent to 3 percent.

3. The tests do not adequately cover the following content areas: Number Systems and Number Theory, Algebra, Statistics, Probability, Geometry, and Measurement.

4. The following process areas are not covered adequately by the tests: Communication, Connections, Reasoning, Problem Solving, and Patterns and Functions.

5. The tests place too much emphasis on procedures and not enough emphasis on concepts.

PART 2: THE FOLLOW-UP STUDY

The aim of the follow-up study (Romberg, Wilson, & Chavarria, 1990) was to demonstrate the existence of test items that are more closely aligned with the *Standards* than are the items found in the six tests of the first study. The investigation drew upon items and tests from two sources: newly developed state tests and foreign tests. The study looked at materials from California, Connecticut, South Carolina, Massachusetts, and Vermont; then it considered materials from several foreign countries—primarily Britain, but also Australia, France, Korea, The Netherlands, and Norway.

The conclusion of the investigation was that there are test items which are currently in use that are more closely aligned with the *Standards* than the six standardized tests that are most widely used now at the eighth-grade level in the United States. Several states are implementing reforms in their assessment practices and have developed tests to reflect the objectives described in the *Standards*. In addition, many tests and test items that are currently used in foreign countries, most notably Britain, surpass American standardized tests in their alignment with the *Standards*. The feature shared by all of these tests and test items is that they are open response, not multiple choice, in format. The content and processes measured in these items are rich and varied. Many of the items are able to assess higher-order thinking with greater ease than do typical multiple-choice questions. Process areas such as Problem Solving, Reasoning, and Communication lend themselves to an open-response format, and this has been borne out in our investigations. In United States tests, only 1 percent of the items could be classified as Problem Solving, 20 percent as Communication (and that at the lowest levels of communication), and 1 percent as Reasoning. In contrast, the open-response tests being developed in several states and in Britain contained excellent examples of items in those three process areas.

The following are examples of test items found on either state or foreign tests:

1. *Connecticut Common Core of Learning Assessment Project*
    Students are given an article from the newspaper entitled, "Survey Finds Many Below Town's Mean Income." They are then given the following questions: Use the article and your understanding of statistics to complete the following tasks:
    1. Write an expository paragraph that begins with either:
        — The headline is fine because . . . OR
        — The headline is absurd because . . .
    2. Write an expository paragraph that begins with either:
        — The article makes sense and has no statistical errors because . . . OR
        — The article is absurd and makes statistical errors because . . .
    3. How can more than half the people be below the mean income?

4. Create a data set to show how more than half the numbers are below the mean. Describe your reasoning.

SURVEY FINDS MANY BELOW TOWN'S MEAN INCOME*

OLD SAYBROOK - A recent survey shows more than half of the respondents earn well below the town's mean annual income of $37,500. Vicki McCourt, a member of the Old Saybrook Affordable Housing Task Force, said 65% of the 200 respondents reported earning less than $30,000 a year. "There are no houses in Old Saybrook that anyone can afford within the mean income—they just cannot do it," McCourt said.

This item is a good example of a problem in the content area of Statistics, the process area of Communication, and a conceptual level of knowledge.

2. *California Assessment Program*
   James knows that half of the students from his school are accepted at the public university nearby. Also, half are accepted at the local private college. James thinks that this adds up to 100 percent, so he will surely be accepted at one or the other institution. Explain why James may be wrong. If possible, use a diagram in your explanation.

This item taps into the critical process areas of reasoning and communication.

The following three items are taken from British tests:

3. *London & East Anglian Group for GCSE Examinations*
   An air-mail letter to India costs 34p. How can you pay correct postage using only 4p stamps and 11p stamps?

This item, in the content area of Number Systems and Number Theory, is a computation problem, but at a conceptual rather than procedural level.

*Source: Connecticut State Department of Education, Connecticut Common Core of Learning Performance Assessment Project. Used with permission. This item has been replaced by a revision pilot tested in Spring, 1991. The project was funded by a grant from the National Science Foundation.

4. "Beefo Cubes" are 2 cm × 2 cm × 2 cm.

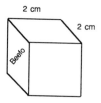

They are sold in a thin cardboard sleeve 4 cm × 4 cm × 8 cm.

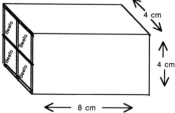

How many cubes are in one full sleeve?

This item could be classified into content areas of either Geometry or Measurement, and a process level of Connections or Problem Solving; it requires a conceptual level of knowledge.

5. *Northern Examining Association*
   The picture shows a woman of average height standing next to a lamp post.
   a)   Estimate the height of the lamp post.
   b)   Explain how you got your answer.

This item is a refreshingly different estimation problem, one that does not involve simply rounding numbers. The "b" part of the problem makes it a good communication problem also. It could be classified as either Measurement or Number Relations and again is at a conceptual level.

6. *The Netherlands*

The results of two classes of a math-test are presented in a stem-leaf-display:

CLASS A		CLASS B
7	1	
7	2	34
4	3	
55	4	
4	5	
1	6	5
1	7	12344668
9966555	8	114
97	9	1

Does this table suffice to judge which class performed best?

Source: de Lange, van Reeuwijk, Burrill, & Romberg (in press). Used with permission.

This open-ended item requires that the student have a conceptual understanding of the data represented in the display and be able to communicate those concepts by means of a valid mathematical argument.

These six items are a sample of the kinds of problems that are possible when one is not bound by the multiple-choice format. Each problem is rich, engaging, and interesting. Tests that are comprised of items such as these can provide a more valid means of assessing the content areas, processes, and levels of knowledge described in the *Standards*. Perhaps most important, a student who encounters test items such as these will come away from the test having *learned* some mathematics through the experience.

SUMMARY

This study has attempted to provide evidence that the six standardized tests most often used by states in their mandated testing programs are not appropriate instruments for assessing the content areas, processes, and levels of thinking called for in the NCTM *Standards*. These tests are "based on different views of what knowing and learning mathematics means" (NCTM, 1989, p. 191). As the *Standards* become more widely implemented in the schools, standardized tests used in the schools will have to change to more accurately reflect the new vision of the mathematics curriculum which the *Standards* outlines. Romberg, Wilson, and Chavarria's 1990 study of state and foreign tests found that there are examples of tests and test items, all of which are open-response in format, that can provide more valid means of assessing the mathematics of the *Standards*.

5

State Assessment Test
Development Procedures

James Braswell

The primary purpose of this chapter is to describe how tests
are developed for state assessment programs. The methods
described are based in part on recent discussions with state
department of education staff in states judged to be representative of a range of approaches to test development. I spoke
with assessment representatives in Florida, Louisiana, Massachusetts, Michigan, and New Jersey. Occasionally, there will
be observations in the chapter that reflect previous experience
with other state testing programs and current work with the
National Assessment of Educational Progress (NAEP) test development team for the 1990 Mathematics Assessment. What
is reported in this chapter is primarily descriptive rather than
evaluative.

The primary purpose of state assessment programs is to
monitor trends in achievement. In some states, such as California (see Chapter 6), the emphasis is on group assessment as
opposed to individual student assessment. In other states, such
as Michigan, the emphasis is on individual student assessment. However, Michigan also provides school and district level
reports that can be used for curriculum evaluation and accountability purposes.

When group assessment is the goal, the larger test instrument can be viewed as a compilation of several shorter test
instruments in which no student takes more than one of the

shorter instruments. For example, the test specifications and evaluation design may be developed into a pool of 400 questions, which are placed on ten 40-item tests. When administered, group data are available on the entire pool of 400 questions, even though any given student only responds to a single 40-item test. When individual student assessment is a major goal, students at a given grade level generally take the same test.

If the primary goal of assessment is improvement in education, test instruments should be designed to provide an appropriate target for instruction. Shavelson (1990) reminds us that what is monitored is what gets taught and that what gets taught has a better chance of being learned than what is not taught. He goes on to point out that:

> By creating achievement tests that more closely measure valued education outcomes, achievement indicators will achieve a reporting function and yet benefit, not create dysfunctions in, education systems. Put another way, achievement indicators should be created in such a way that if education institutions teach to the test, they will be teaching what is important for students to know. (Shavelson, 1990, p. 7)

My overall impression is that the states engaged in assessment activities do a thorough job of developing specifications. Advice and review is sought from a wide spectrum of the educational community. Increasingly, states seem aware of the fact that their assessment programs should not be narrow in scope, but rather reflect the range of desired instructional outcomes. Massachusetts, for example, has introduced open-ended sections in their state assessment program and has published a series of booklets to describe the implications for instruction. The recently published NCTM *Curriculum and Evaluation Standards for School Mathematics* (1989) is a useful document for states to consider as a meter stick against which the current range of their assessment objectives can be measured. The implications of these standards are discussed in Chapter 3.

Although some states use commercially developed, norm-referenced tests as a component of state assessment, other states have legislative mandates to develop instruments that are tailored to local objectives and that involve teachers and other educators in the state. States that opt for using an existing norm-referenced instrument generally select an instrument that represents the best fit with the state's objectives. Although such tests are likely to contain a number of items that state department staff and local educators view as inappropriate, budget and other considerations can make the use of commercially available assessment instruments appealing. Rather than contracting with others to develop instruments tailored to the state's objectives, items in "best fit" commercially available tests can be matched with local objectives. For those areas deemed important, but not included on the test of best fit, the state can develop, or contract to have developed, a tailored instrument to fill in gaps. This possibility will not be explored further, but it does represent an alternative to a completely tailored program. Chapter 4 of this book examines a number of standardized tests. Based on the findings reported, one might expect to find a rather large gap between the current content of these tests and what is called for by the NCTM *Standards.*

State mandated testing has increased substantially in recent years. As Coley and Goertz (1990) report:

> In 1989–90, 47 states required that local school districts test students at some point(s) between grades 1 and 12, an increase of five states since 1984–85. Thirty-nine of those states test students using state-developed, state-selected, or state-approved tests and assess student performance against state-established performance standards. (p. 3)

Statewide assessment is commonly conducted at three grade levels, such as Grades 4, 7, and 11. Increasingly, there is a high school graduation component administered at one or more levels in Grades 9–12. In 1990, twenty states required that students pass a basic skills or other competency assessment before receiving a high school diploma.

The major steps in developing any test are the following:

- setting specifications
- writing items
- reviewing items
- field testing items
- test assembly
- test review

The pages that follow will describe in a collective way how certain of the states accomplish the above steps. Then, three case studies will be presented to give the reader a better picture of how particular states approach test development. Finally, there is a section on test development activities related to the 1990 National Assessment in Mathematics.

Test Specifications

Most states have been conducting state assessments for several years and use specifications that have evolved over time. Typically, states actively involve classroom teachers, curriculum coordinators, and others in setting and reviewing specifications. It is becoming increasingly common for states to seek the advice of those outside the immediate educational community—e.g., representatives of business and industry—in the formulation of broad educational goals. Such goals serve to focus the assessment activity and provide a support base.

In some states, minimum performance standards have been set for the schools by the state, and test specifications are developed by local test-writing committees to reflect these standards. State board of education staff may review and approve the specifications. These specifications are sometimes too general to guide item writers, and states may contract with an independent contractor to develop item-level specifications and one or more sample items. The independent contractors frequently involve teachers and other educators in setting the item-level specifications, which normally undergo extensive review. When consensus has been reached, specifications and sample items may be printed in a form that can be shared with local school districts. Some states use an eclectic approach in their assessment activities. Objectives developed by others are reviewed and adapted to local goals.

In order to determine which topics are important, some states have prepared a long list of topics for the grade levels being assessed. This list is mailed to teachers throughout the state and they are asked to rate topics. For example, a topic rated 4 might be viewed as extremely important while one rated 1 would be viewed as unimportant. Average ratings are compiled for each topic. These ratings guide those who set specifications for the various tests. Few states, however, follow such an extended consensus-building process. In fact, one could argue that a few well-informed educators are in a better position to determine an appropriate list of skills and topics than might be obtained by means of a statewide survey of teachers in targeted grades.

In addition to content specifications, increased attention is being given to the process dimension. Some states (e.g., see Massachusetts Department of Education, 1987, p. 29) use category specifications recently proposed by the National Council of Teachers of Mathematics. These categories are Procedural Knowledge, Conceptual Understanding, Problem Solving, and Reasoning and Analysis. Other states (e.g., see Michigan case study later in this chapter) use categories that have greater specificity, such as Mental Arithmetic, Estimation, Computation, Conceptualization, Applications, Calculators, and Computers. Whatever process categories are used, it is important that they have two characteristics: (1) They should span the range of what it is important to measure, and (2) What is included under each process dimension needs to be well defined. For example, it is not sufficient to simply list "conceptual understanding" as a category. The category should be described and illustrated so that item writers, reviewers, and users will understand what the categories mean. Note that the NCTM categories are generally disjoint process categories, whereas in the second listing Mental Arithmetic, Estimation, and Computation tend to be procedurally oriented. By specifically including calculators and computers, these tools are singled out as important elements in the assessment framework.

In summary, content specifications for state assessment instruments are generally developed under the direction of state department of education staff with the help of teachers, curriculum coordinators, other educators, and occasionally representatives of business and industry. State department staff almost universally depend on the talents and perspectives of those outside the department. This is especially true of state departments with limited subject-matter expertise on staff, but it is also true of those state departments that have a fairly large staff of subject-matter specialists.

While teachers and curriculum specialists play a major role in setting content specifications, the statistical specifications are generally determined by others. The primary statistical considerations concern the difficulty level of the test and the range of difficulty of the various items that will be used on the test. Setting statistical specifications is usually done jointly by the appropriate state department staff and the independent contractor. Those states that develop their own instruments without the assistance of an independent contractor generally have staff with psychometric training within the state department.

The statistical specifications depend on the purpose of the assessment. In the mastery-level or basic skills programs, which were popular in the 1970s, the statistical specifications generally called for items that the majority of students could pass. In the more comprehensive assessment programs characteristic of the 1980s, in which the goal is to assess a wide range of student achievement, statistical specifications call for a range of difficulty to differentiate among levels of achievement. For the most part, statistical specifications tend to be a function of the content and ability specifications. For example, a procedural knowledge question in a certain content category is usually easier than a conceptual understanding or problem-solving question in the same content domain. Rigid statistical specifications do not appear to drive state assessment programs.

Item-Level Specifications

Some state assessment programs provide few details about the specific characteristics of items that are to be written to meet the content and ability specifications. For a given content specification, a wide range of items would be viewed as acceptable even though some would be considerably more difficult than others. Other states provide extremely detailed item-level specifications. These specifications may describe the characteristics of the numbers to be used and the method of creating distractors. In such cases, item writers have less freedom in posing questions, and items written for a given specification usually tend to be similar in content and difficulty. It is generally easier to pinpoint instructional deficiencies when performance on such items is weak. A drawback of such specificity is that the items so written tend to test lower-level cognitive skills at the expense of the higher-order problem-solving skills that are viewed as increasingly important by mathematics educators. While highly detailed specifications are an advantage in identifying deficiencies in skill areas, highly specified item parameters may impact instruction negatively if there is pressure on schools to do well on these items at the expense of other important instructional outcomes.

Some state assessment programs are beginning to include open-ended items which require students to explain, construct, measure, graph, analyze, and compute. Such exercises are more expensive to administer and score, but they provide more definitive and interesting information about what students can and cannot do. Those states that administered NAEP blocks in Grades 4, 8, and 12 of the 1990 assessment found traditional multiple-choice questions as well as a wide range of open-ended and calculator-active items. During testing, four-function calculators were furnished in Grade 4 and scientific calculators were furnished in Grades 8 and 12.

Item Writing and Review

Once the content and ability-level specifications have been determined, the next step is to write items which meet these specifications. States use different approaches in item development. In some states items are written by an independent contractor who sometimes involves teachers in the target state. Other states appoint subject area committees to write the questions. As a rule, these committees consist of teachers, curriculum specialists, and other educators with knowledge of the curriculum at the relevant grade level(s). Still other states contract with state universities to develop items to fit the specifications. Several states use NAEP items or blocks of items that have been written to be broadly applicable to curricula across the nation and to be consistent with current thinking among mathematics educators. (See section on NAEP item-development procedures for more information on this activity.)

Item review procedures also differ from state to state. Usually, items written by a contracting agency are reviewed by staff at the agency and edited for style, clarity, and grammar. They are then reviewed by state department staff. It is fairly common for the state to appoint a committee consisting of teachers, curriculum coordinators, and/or independent consultants to review items. These reviews attempt to answer several important questions:

1.   Is the item technically correct (i.e., does it have one and only one correct answer, is it unambiguous, and is it grammatically correct)?

2.   Does the item meet the specification?
3.   Is the item written at a level appropriate for the target grade?
4.   Are the distractors reasonably chosen?

At the review stage, items may be rejected, accepted, or revised. The end product of the item-writing and review phase is a collection of questions ready for field testing. As part of the review process, items are usually classified according to a content-ability/process matrix, such as the one below. If gaps in coverage are observed—i.e., there are too few items in certain specification categories—additional writing and review may be required.

Table 5–1
Content-Ability Matrix

		Content Area				
Mathematical Ability		Numbers & Operations	Measurement	Geometry	Data Analysis, Statistics, & Probability	Algebra & Functions
Conceptual Understanding						
Procedural Knowledge						
Problem Solving						

An example of a question that could go in the cell corresponding to "Conceptual Understanding, Geometry," is the following:

Points P and Q are on opposite sides of a rectangle that has length 10 and perimeter 32. If x represents the distance between P and Q, what is the LEAST possible value of x?

(A) 4   (B) 6   (C) 8   (D) 10   (E) 16

Correct answer (B).

Item Tryout

States usually field test items prior to their inclusion on actual tests. In most state assessment programs, the independent contractor assembles the tryout blocks. Tryout procedures vary, but generally local school districts serve as pilot sites. Some states pilot items in experimental forms along with the operational assessment. Items may be in separate blocks or embedded in operational test forms. Other states pilot items independent of operational administration and may provide feedback to participating districts. Sample sizes are usually fairly large. For example, Florida field tests items on approximately fifteen hundred students. When items are tried out on sufficiently large samples that are representative of the intended population, one can be reasonably sure that tryout data will hold up on operational administrations.

Test Assembly

All items involved in the tryout phase are candidates for operational forms. As a result of the field testing, some items are judged as unsatisfactory and are not placed in the item pool from which operational forms are assembled. States view field testing as an opportunity to replenish and expand their item pools and to fill gaps that have been identified by those involved in the assessment effort. Normally, item pools contain items tried out at different times, some of which may have been previously used in operational assessments.

In some states the independent contractor assembles and produces operational forms of the test according to the specifications established before item development began. In other states, state department staff assemble the operational forms.

Usually, new operational forms of the test are reviewed by a variety of people outside the state department. Such reviews are in addition to the review steps taken by the independent contractor. In Massachusetts, for example, a committee consisting of teachers and curriculum coordinators, as well as an equity review committee, review operational tests. In Michigan, tests assembled by an independent contractor are reviewed by a committee consisting of teachers, curriculum coordinators, and college faculty. Michigan Department of Education staff and two independent consultants also conduct reviews.

Not all states equate new operational forms to those built in previous years. States that do not are likely to have substantial item overlap from one form to the next. For example, in Massachusetts only about 15 to 20 percent of the questions are replaced from one year to the next. Performance on the remaining items serves as a basis for making comparisons. In Florida,

test assemblers strive to build comparable forms based on statistical data available from previous administrations and item
tryouts. In Louisiana, the blueprints for test and item specifications at Grades 3, 5, and 7 are very detailed. New forms are
assembled using a Rasch model and equated to previously administered forms. A test built to such a model uses item difficulty parameters that will provide the most accurate score information at those score levels which are viewed as the most
important. Technically, the standard error of measurement is
controlled at various points on the reporting scale.

Graduation Tests

Some states now have a test requirement for high school graduation (that must be passed in order to graduate from high
school). In Florida, for example, the graduation test is first
administered in Grade 10 and is based on skills approved by
the state board of education. The test contains two sections,
communications and mathematics. Each section contains seventy-five items measuring fifteen skills. Items are selected within
each skill area on the basis of Rasch difficulty values. Test
scores are equated to the base-year scale on the basis of common items sets and a linking design.

In New Jersey, the state department of education has discontinued the Minimum Basic Skills Tests that were administered during the 1980s in Grades 4, 7, 9, and 11. In 1985, the
state implemented the High School Proficiency Test (HSPT),
which students may take beginning in Grade 9. The HSPT
applies to students who enter high school in September 1985,
or thereafter. In order to graduate from high school, students
must achieve a passing score on the HSPT sometime between
Grades 9 and 12.

Specifications for the HSPT are developed by state committees of educators and business people. Based on these specifications, item development committees, consisting of teachers,
curriculum supervisors and others, write sample items to fit
the specifications. The specifications, together with sample items,
go to an independent contractor who writes additional items for
each specification. These items are returned to the item development committees for review and revision. The committees
decide whether the items meet specifications and make appropriate revisions. Items judged appropriate are returned to the
independent contractor for field testing. Items are placed in

tryout tests and field tested in ninth-grade classes throughout the state. Following the field testing, item data is available and the committees rate items as either acceptable or not acceptable. The state department of education chooses from among the acceptable items those which satisfy the test specifications and returns them to the contractor. The contractor is responsible for producing test copy which is reviewed by state department staff. A minority review group reviews the initial items that are written for field testing as well as the final test.

Within a given skill specification, the HSPT content can vary from year to year. The concept of average, for example, might be tested in a straightforward way one year and in a somewhat more advanced way in a subsequent year. A straightforward approach might require the student to simply find the average of a given set of numbers. A more demanding exercise might require the student to solve an exercise such as the following:

The average of five numbers is 18. If one of the numbers is 10, what is the average of the other four numbers?

(A) 2   (B) 14   (C) 20   (D) 24   (E) 26

The New Jersey state department is currently planning to develop forms of the examination that are not immediately released so that equating can be done more directly. In 1993, the state plans to shift to an eleventh-grade test that will be more demanding than the present ninth-grade test, which contains material through pre-algebra and some geometry. An early warning test may be offered in Grade 8.

CASE STUDIES

In order to provide a clearer idea of how particular states develop their assessment instruments, the approaches taken by Florida, Massachusetts, and Michigan are described below.[1]

---

[1] The author wishes to express his thanks to Mark Heidorn (Florida), Elizabeth Badger (Massachusetts), and Sue Rigney (Michigan) for providing information about their state assessment activities and for reviewing a draft of the descriptions provided for their respective states. Rebecca Christian (Louisiana) and Stan Rabinowitz (New Jersey) also provided helpful information about current assessment activities in their states.

Case Study — Florida
 1. *Who develops and reviews specifications?*
    Minimum Student Performance Standards are devel-
    oped under the direction of the state department of
    education by local writing committees representing
    Florida schools. The standards are defined by compo-
    nent skills, which are the instructional objectives as-
    sessed by the program. These standards and skills
    are reviewed and approved by the state board of edu-
    cation. The next step is to develop item-level specifi-
    cations. These are generally written under contract by
    state university centers or staff together with person-
    nel from local school districts. The contracting agency,
    together with school personnel, are involved in writing
    and reviewing these specifications. Typically, a team
    writes both a specification and a single sample item.
    After this phase, the state conducts a department-
    level external review with a group of teachers and
    curriculum leaders who are experienced with the con-
    tent the specifications are designed to measure and
    who have been selected to represent major ethnic
    groups and geographical regions of the state. After
    the item-level specifications and sample items have
    been reviewed, they are sent to the sixty-seven school
    districts for approval and validation. The goal is to
    have each district conduct a thorough review. Most
    districts take this task seriously. If some specifica-
    tions need to be reworked as a result of these reviews,
    they may be resubmitted to the districts for approval.
    Finally, item specifications and sample items are then
    prepared in a formal document which is shared with
    the districts.
 2. *Who writes items?*
    The process for writing items is similar to the item
    specifications and sample items process. The state
    contracts again with state university centers or staff
    or with local school districts who obtain teachers at
    the district level to write questions. All test items,
    after having been internally reviewed by the contract-
    ing team, are pilot tested by the contractor in Florida

classrooms. As a part of pilot testing, students are interviewed about their solutions to questions.

3. *Who reviews items?*

   The items are reviewed by the contracting agency at the time they are written. This review involves teachers employed by the contracting agency.

4. *Who edits items?*

   Editing is conducted by the contractor in several phases. Items are also edited by department staff as part of the review process.

5. *Are items tried out?*

   Items are initially tried out on groups of twenty to twenty-five students. Following this phase, the contractor revises the items and prepares them for state department review. After this review, the items are field tested statewide, using samples of approximately fifteen hundred students. The field testing is done at the time of the regular assessment. Usually, the experimental forms are given immediately after the regular assessment is completed. In some instances, items are embedded in operational test forms.

6. *What is the item selection/rejection process?*

   Questions may be revised or rejected at any stage of development. Questions that prove to be satisfactory as judged by the field test results are placed in the state department's item bank.

7. *Who assembles the final test?*

   Members of the state department staff assemble new forms of the test using questions in the item bank. These banks are used for Grades 3, 5, 8, and 10. Skills assessed in a given year may vary from the previous year, but a substantial core of common items is generally used. Although the item specifications may be the same, the items used to measure those specifications might change. State department staff generally attempts to select questions that have approximately equal percent-correct values to measure the same specification. This tends to assure that tests are roughly comparable in difficulty from year to year. In a given year, approximately 60 percent of the items may be identical to those used in a previous year. The remainder test the same specification, but with a different question. A small number of items pertaining to certain skills may be removed from or added to the test each year.

8. *What are the review steps?*
The test is reviewed internally by state department staff. One staff member has primary responsibility for this process and several others are involved in assembly and review.

9. *Are forms equated from year to year?*
In Grades 3, 5, and 8 the test assemblers try to build comparable forms based on available statistical data. Staff replace approximately 30 to 40 percent of the items each year. The graduation test, which is first administered in Grade 10, is based on skills approved by the state board of education. This consists of two sections, communications and mathematics. Each section contains seventy-five items measuring fifteen skills. Items are selected within each skill on the basis of Rasch difficulty values. Test scores are equated to the base year scale on the basis of common item sets and a linking design.

10. *What use is made of the tests?*
Individual school districts receive considerable feedback at each grade level. The following materials are made available: individual student reports, class summaries, school summaries, district and regional summaries. These reports enable one to identify individual students and the percentage of students at each level who fail to master specific skills and standards. If the average mastery rate for a school falls below a certain level, the school is considered to be deficient. Students who do not pass must be given remediation, but they are not held back from the next grade on the basis of the test score. Decisions to retain students could be made on the basis of the results of remediation.

11. *Miscellaneous comments.*
Part I of the assessment program at Grade 10 consists of basic skills. Part II is an assessment of the

application of the basic skills to every day situations. Part II was moved to the tenth-grade so that students who fail can have additional opportunities to take the test during their junior and senior years. Florida is also developing a series of secondary school subject area tests—English I Skills, English I, English Honors, Algebra I, Algebra I Honors, Introduction to American History, American History, and Advanced American History.

Case Study — Massachusetts

1. *Who develops and reviews specifications?*

The Massachusetts statewide assessment program uses the NAEP objectives, but adapts them to fit local goals. The assessment takes a forward-looking approach as opposed to evaluating the status quo. The assessment covers Grades 4, 8 and 12. The focus is not just on what is taught at these specific grade levels, but on what it is important for students to know at each of these grade levels.

In order to determine what topics are important, all public schools in Massachusetts are surveyed periodically. Teachers are provided with a list of content topics and are asked to rate the topics according to their importance at the three grade levels. The state appoints a committee in each subject area consisting of Massachusetts public school teachers and curriculum coordinators. This committee interprets the ratings of the various topics and sets the specifications for assessment at each grade level. A booklet providing a framework for the assessment is prepared. The booklet contains a description of the content to be assessed in Grades 4, 8, and 12. Sample questions are also provided. A two-dimensional assessment matrix provides guidelines for writing items and assembling the final test. In addition to content areas, such as Numbers and Numeration, Measurement and Geometry, Problem Solving, and Probability and Statistics, there are four process categories:

- Procedural knowledge
- Conceptual understanding

- Problem solving
- Reasoning and analysis

The process categories, the content categories, and many content subcategories become the reporting categories for school results.

Approximately two to three hundred questions are used at each grade level. There are at least twelve questions in each reporting category; for example, in Operations: Whole Numbers, there would be at least twelve questions scattered across the four process categories.

2. *Who writes items?*

Questions are written by an independent contractor. Massachusetts also uses questions from NAEP and other sources.

3. *Who reviews items?*

Items are reviewed by the independent contractor, committees consisting of teachers and curriculum coordinators, and the Massachusetts state department staff. An equity review committee addresses minority and gender issues. In the future, equity issues will be addressed by a curriculum advisory committee.

4. *Who edits items?*

Items are edited by the independent contractor and by various reviewers along the way.

5. *Are items tried out?*

All new items are tried out in Massachusetts public schools. The state department identifies the schools and assists in piloting. The independent contractor prepares the tryout tests. All questions that appear in tryout tests are reviewed and approved by the committee of teachers and curriculum coordinators. Questions used from NAEP are not tried out since these have already been through a similar process.

6. *What is the item selection/rejection process?*

Newly written questions as well as questions from other sources are considered for the final assessment. Individual items are rated by the committee and placed in one of three categories:

1. The committee likes the question, and it is appropriate for assessment.
2. The committee likes the question, but some modification is needed.
3. The question is not appropriate and is rejected.

The committee reviews, culls, and classifies questions, and matches them with the content-process matrix. If there are gaps, these are filled in by the independent contractor.

7. *Who assembles the final test?*

The committee approves all items that go into the final test. The pool of approved items then goes to the independent contractor who assembles booklets according to a matrix-sampling plan that will meet the content and process goals of the assessment. The purpose of the assessment is not to provide a report of individual student performance, but rather to provide data at the school, district, and state levels about the performance of Massachusetts students on important content and process dimensions. Some open-ended questions are included in the assessment.

8. *What are the review steps?*

The test is reviewed by a committee of Massachusetts teachers and curriculum coordinators, Massachusetts state department staff, and the independent contractor.

9. *Are forms equated from year to year?*

Although test forms are not equated from year to year, results can be compared, since only about 15 to 20 percent of the questions are usually replaced between consecutive years.

10. *What use is made of the tests?*

The test results are used in three basic ways:

- At the school level, reports are prepared if more than twenty students in a school are tested at a given grade level.
- Detailed reports are prepared for use by the state department of education.
- Questionnaires administered to teachers, students, and principals are used to help explain the results

of the assessment. For example, actual test questions are shown on the teacher questionnaire and teachers are asked about the students' opportunity to learn what is tested by the various questions. This type of information helps interpret results.

11.   *Miscellaneous comments.*

A primary goal of the Massachusetts assessment program is to help schools evaluate and improve their performance. Several activities currently underway are directed toward this goal. For example, responses to the open-ended sections of the test have been analyzed by curriculum committees and the items, results, and implications for instruction will be published in a series of booklets. Teachers will be encouraged to administer the questions to their own students for comparison and diagnosis.

Massachusetts is also promoting performance testing in mathematics and science. Approximately seventy teachers have been trained to test a random sample of one thousand students in Grades 4 and 8. Videos of the testing, together with written reports of the results, will be used to help teachers improve classroom assessment.

Case Study — Michigan

1.   *Who develops and reviews specifications?*

The Michigan Educational Assessment Program (MEAP) tests all students in mathematics and reading at Grades 4, 7, and 10. The basis for these tests are the "Essential Goals and Objectives in Mathematics" (developed by groups of Michigan content specialists and approved by the state board of education) and item specifications. State department staff are currently revising the specifications for the mathematics tests as part of a comprehensive test revision process resulting from the recent approval of new "Essential Goals and Objectives in Mathematics." The first every-pupil MEAP administration of the revised mathematics tests is scheduled for 1991.

The "Essential Goals and Objectives" may be viewed as a table of specifications for the test, whereas item specifications detail the standards and conventions to be employed within individual items. Item specifications for the current mathematics tests were developed by the Michigan Council of Teachers of Mathematics several years ago. Some gaps in those item specifications have been identified and will be corrected as part of the test-revision process.

The test-revision process includes consultation with content area specialists (teachers, curriculum coordinators, and others knowledgeable about the mathematics curriculum) throughout the state. A twenty-member advisory group, the Mathematics Coordinating Committee, has been formed to advise the MEAP staff on test-related issues, including item writing. The committee represents different geographical regions of the state as well as different vertical perspectives—that is, college faculty, teachers at appropriate grade levels, and curriculum specialists. Such diversity tends to assure a good balance of viewpoints. Also, this group tends to be a good resource for disseminating accurate information about the assessment program across the state.

2.   *Who writes items?*
     Items are generated by mathematics educators in Michigan based on objectives approved by the state board of education. Meetings are held at which math educators establish a conceptual base for formulating questions. Important instructional outcomes are identified and the questions developed. The items are written by various subgroups for each of several content strands. For example, the content strands include areas such as fractions, decimals, and geometry. The writing groups also focus on the process dimension. The following process categories are used: conceptualization, mental arithmetic, estimation, computation, applications, calculators, and computers.

3.   *Who reviews items?*
     Items are reviewed and edited by an independent contractor. The contractor comments on (a) the match between item and specification, (b) item classification, and (c) duplication of items and psychometric considerations. The contractor types items, supplies relevant art work, provides comments on individual items, and

then returns them to the original writing group for a strand review. Classroom teachers and content specialists who wrote the items review them along with the contractor's comments. Substantial modification may occur at this stage. Items are also subjected to in-house review by one member of the MEAP staff and by two consultants. The following questions are asked as each item is reviewed:

1. Is the item in its current form appropriate to the objective?
2. Is the item clearly worded?
3. Is the item grade-level appropriate?
4. Are the distractors reasonable?
5. Is the artwork correct?

4. *Who edits items?*
   Items are edited by the independent contractor (see step 3 above).

5. *Are items tried out?*
   The most recent item tryout was informal, consisting of questions that were administered to a sample of students from representative districts. Following this phase, MEAP staff asked that schools volunteer for a formal item tryout. Districts agreed to provide students, and test results were shared with the pilot schools. Michigan also pilots the actual test to see how the questions fit together as a group and to provide a trial run of administration procedures and support materials. The next test pilot is scheduled for the fall of 1990.

6. *What is the item selection/rejection process?*
   The twenty-member advisory group and various other review groups have input throughout the test development process. The final selection of items is made by MEAP staff.

7. *Who assembles the final test?*
   The final test is assembled by the independent contractor with a detailed recipe furnished by MAEP. Attention is given to appropriate content areas as well as to the problem-solving process dimension.

8. *What are the review steps?*

   The test prepared by the independent contractor is reviewed by the twenty-member committee, MAEP staff, and two consultants. There is a pilot trial of the test as described under Step 5 above. Final test copy is proofed by MEAP staff, the advisory committee members, and content specialists.

9. *Are forms equated from year to year?*

   In the past, the test changed very little from year to year and there was no reason to equate. New forms are being developed, and they will be equated and year-to-year trends charted.

10. *What use is made of the tests?*

    Test results are used to generate student reports, school and district level reports, and detailed state summaries. At the school level, results provide useful information for:

    - individual student remediation,
    - curriculum review and improvement, and
    - students, parents, school boards, and the public.

    At the state level, results are also used as a basis for funding allocation and research.

11. *Miscellaneous comments.*

    Michigan is considering the development of an Employability Skills assessment. The impetus for this activity is the perceived lack of certain basic skills on the part of those who enter the job market.

National Assessment of Educational Progress (NAEP) Procedures

Because so many states are involved in various aspects of national assessment, including the use of NAEP objectives, items, and comparative data, it may be useful to outline briefly NAEP development procedures. In many ways, the NAEP procedures have parallels in state assessment efforts. However, the consensus-building activity, specifications-setting process, and test-development procedures are generally more complex. An outline of the key NAEP activities leading to the 1990 assessment is provided below.

Planning Phase

In 1988 Congress authorized NAEP to provide for voluntary state-by-state assessment in addition to the traditional assessment activities carried out by NAEP over the last twenty years. Government funds were made available to the Council of Chief State School Officers (CCSSO) to lay the groundwork for state comparisons. The charge to the CCSSO was to recommend objectives for the state-level assessment and to suggest how state results should be reported. A planning group under the direction of the CCSSO was appointed to develop objectives for Grades 4, 8, and 12. Legislation was subsequently passed to specify that Grade 8 would be the target grade for state-by-state assessment activity. Because the 1990 assessment is designed to provide state-level performance reports as well as national reports, in its planning phase careful attention was given to the objectives of the various states. Also, advice was sought from various other groups.

The objectives developed under the direction of the CCSSO were then formulated and refined by a Mathematics Objectives Committee composed of teachers, administrators, mathematics educators from various states, mathematicians, parents, and citizens. Sample questions were also developed. These materials underwent extensive review by the states, by NAEP policy groups, and by others. The objectives underwent further review by NAEP's Item Development Panel and a framework for the assessment was published in November 1988.[2] The framework called for assessment in five content areas and three ability levels. The content areas are Numbers and Operations; Measurement; Geometry; Data Analysis, Statistics, and Probability; and Algebra and Functions. The ability levels are Conceptual Understanding (CU), Procedural Knowledge (PK), and Problem Solving (PS). Percentages were set at each of Grades 4, 8, and 12 for both content areas and ability levels. For example, at Grade 8 the percentages for CU, PK, and PS were 40 percent, 30 percent, and 30 percent, respectively. The Mathematics Objectives booklet provides a detailed description of what the abil-

---

[2] This publication, *Mathematics Objectives 1990 Assessment*, can be ordered from the National Assessment of Educational Progress at Educational Testing Service, Rosedale Road, Princeton, NJ 08541–0001. Refer to publication No. 21–M–10.

ity levels encompass and the range of subtopics included under the various content areas. Sample items are also included.

The ten-member Mathematics Item Development Committee then began work on the item-writing phase of the project. In addition to this group, about twenty-five other mathematics teachers and educators across the country were asked to develop new items. Test specialists at the Educational Testing Service (ETS) conducted an on-site training session for thirteen of the item writers and made specific item-writing assignments based on the objectives.

The following steps were then taken by the NAEP/ETS development team:

1. Internal review of newly developed items. Items were reviewed for clarity, appropriateness to the specification, technical accuracy, age-level appropriateness, distractors, and for possible offensiveness to population subgroups. Gaps in specifications were filled by test development staff.

2. Items that were judged acceptable were classified and filed. Rejected items were filed separately.

3. Seven mathematics test specialists assembled draft field test blocks, keeping in mind the overall target specifications for the 1990 assessment, as well as making judgments about the overall content and difficulty of individual blocks.

4. Individual blocks were reviewed internally by other test specialists, the first reviewer working with the assembler to achieve mutually acceptable revisions or replacements.

5. A second reviewer then reviewed the revised version to yield a final draft of the block.

6. All draft blocks were typed and submitted to several program staff for their review.

7. Final revisions were made to blocks which were then proofed and keyed by yet another test specialist.

8. The eighth-grade blocks, which will be used for the state trial assessment, were reviewed by about sixty state representatives from forty states. This group consisted primarily of mathematics special-

ists and testing directors. The blocks were reviewed in groups consisting of ten state representatives, one NAEP/ETS staff member, one ETS test development specialist, and one staff member from the National Center for Educational Statistics (NCES). This process assured that each block was reviewed independently by two groups. Suggestions from the various groups regarding the appropriateness of questions, wording, classification, and other issues were collated and shared with the Item Development Committee, which met shortly thereafter.

9. After all blocks were prepared, they were mailed to members of the Item Development Committee for review. A committee meeting was subsequently held to discuss and revise the questions so that final blocks could be prepared for field trials. At this meeting, suggestions made by the sixty state representatives were evaluated and incorporated into the blocks, or rejected for use.

10. After the Item Development Committee meeting, revisions were made to blocks by the various test assemblers, and each block was edited for clarity, usage, and format. All blocks received a sensitivity review as prescribed by ETS guidelines.

11. Final draft blocks were submitted to NCES for review and clearance by the Office of Management and Budget.

12. Camera-ready copy of each block was then prepared and reviewed by the original test assembler and another mathematics test specialist.

13. Galleys were produced and following additional quality control checks by test specialists and NAEP program staff, the materials were printed and shipped to field test sites.

14. In February 1989, field trials took place in the nation's schools.

15. Assembly of the final 1990 assessment blocks followed much the same development and review procedures that are outlined above. However, item statistics on multiple-choice and open-ended items,

as well as the readability characteristics of student solutions to the open-ended items, were available to assist in final assembly. Also, items from previous NAEP assessments together with the newly field-tested items were available for final assembly. All items were classified and assemblers of final blocks maintained counts of items selected so that overall content, ability, and statistical specifications were met.

SUMMARY

My overall impression is that the states engaged in assessment activities do a thorough job of developing specifications. Advice and review is sought from a wide spectrum of the educational community. Also, representatives of business and industry are sometimes involved.

Items and tests appear to receive thorough reviews. The one area that seems to need more attention is item writing. No matter how carefully specifications are set, if the questions written to fit these specifications are not crafted with skill and care, the impact of the assessment will not be as significant as it might otherwise be. The new NCTM *Standards* are beginning to have an impact on state assessment activities. In the future, the *Standards* should play an even greater role in providing proper focus for test content and the process dimension along which content resides. At the present time, states are making strides in extending the range of processes measured to better cover conceptual understanding and higher-order thinking. However, in most available assessment instruments there are many more examples involving standard procedures and simple understandings than exercises that call for deeper understanding and significant problem solving.

It might possibly be helpful to view each question written and selected for inclusion on the assessment instrument as a candidate for the front page of the *New York Times*—displayed there so all can see what it is important to test. Viewed in this context, certain questions might be withdrawn from consideration.

6

Test Development Profile of a State-Mandated Large-Scale Assessment Instrument in Mathematics

Tej Pandey

From the methodological point of view, the character and design of test instruments that are optimum for individual-level and for group-level assessments are quite different. This paper examines the nature and design of test instruments of a large-scale assessment program, the California Assessment Program (CAP), which is designed to provide reliable group-level information. The paper also describes the test development process as it has evolved over a period of fifteen years to meet the curriculum demands of the time.

Large-scale assessments can be classified into two main types: in one the interest lies primarily at the level of *individuals* and in the other the interest lies primarily at the *group* level. Individual-level assessments typically use test information to rank a student on an established norm, find an individual's strengths and weaknesses, and determine whether a student has mastered specific course content. Group-level assessments typically use the information to measure the achievement level of students in a school, district, or regional system for purposes of determining program effectiveness. Group-level assessments generally are concerned with trends in achievement from one cycle of assessment to the next and may even incorporate provisions in assessment designs to relate trends with other factors.

PURPOSES OF LARGE-SCALE ASSESSMENTS

Large-scale assessments generally have their genesis either in federal programs, such as Titles III and V of the National Defense Education Act (NDEA), or in state-level accountability programs. Both federal- and state-level accountability programs require the centralized collection of test scores, primarily from commercially published standardized tests, by state departments of education. In the last two decades, although the character of assessments for accountability has changed and the purposes of assessment have varied from program to program, they are usually designed to evaluate curricular programs, gather curricular and related information for policy development, and stimulate curricular practices. As Cronbach (1980) states:

> Outcomes of instruction are multidimensional, and a satisfactory investigation will map out the effects of the course along these dimensions separately. . . . To agglomerate many types of post-course performance into a single score is a mistake, since failure to achieve one objective is masked by success in another direction. . . . Moreover, since a composite score embodies (and usually conceals) judgments about the importance of the various outcomes, only a report that treats the outcomes separately can be useful to educators who have a different value hierarchy. (p. 236)

In other words, assessment for broader educational and societal uses calls for tests that are comprehensive in breadth and depth. Both breadth and depth can be covered by including a large number of questions for assessment using a variety of assessment modes, such as direct assessment of performance, open-ended questions, and portfolios, in addition to the multiple-choice format.

MULTIPLE-MATRIX SAMPLING

Since the interest in program assessment is not to obtain scores for individual students but to see how well a body of subject matter has been learned by a cohort of students, multiple-matrix sampling or item sampling can be used effectively. Under matrix sampling or an item-sampling plan, a universe of test items is subdivided into multiple test forms with each form administered to a certain number of examinees selected randomly from the population of examinees. Although each examinee is administered only a portion of the test items in the total pool, the results from each subtest may be used to estimate the parameters of the universe scores, such as the mean, variance, and associated standard errors.

Item-sampling procedures have several advantages over the conventional testing procedure. First, since in item sampling no student takes more than a small portion of the total item pool, the test takes less classroom instructional time, is less fatiguing to students, and results in greater cooperation from students and school authorities. Second, since it allows for testing a large number of questions, it results in a comprehensive assessment leading to more content-related information. Third, it produces more reliable group scores. Lord (1962) showed that for a fixed number of student-item confrontations, the group mean of the item domain is estimated most reliably when the size of the item subset is one, that is, when each item is taken by a different sample of students. Greater reliability is achieved because item responses tend to be positively correlated over the population: two items presented to one student will not generally supply as much information about the mean of the item domain as two items presented to different students.

THE CAP'S ITEM-SAMPLING DESIGN

As of spring 1990, the California Assessment Program administers tests in reading, written expression, and mathematics annually in Grades 3, 6, 8, and 12. Also, science and history-social science are assessed at Grade 8, and direct writing is assessed at Grades 8 and 12. The assessment program uses a nonoverlapping item-sampling design (see Pandey, 1974; Pandey and Carlson, 1975, 1983) for student assessment. Since the assessment program provides information to all schools "small and large," all students are tested rather than a sample of students. Table 6–1 shows the total number of questions and the number of forms in each of the content areas for each grade tested.

From Table 6–1 it is apparent that the *Survey of Basic Skills: Grade 3* (1980 version) consists of a total of 1,020 items—

240 in reading, 420 in written expression, and 360 in mathematics—divided into thirty unique test forms. Under the item-sampling procedure, each test form consists of a total of 34 questions made up of 8 questions in reading, 14 questions in written expression, and 12 questions in mathematics. Each form of the test is constructed to have an equal number of easy

Table 6–1
Number of Questions and Test Forms of CAP Tests
at Grades 3, 6, 8, and 12

	Grade 3	Grade 6	Grade 8	Grade 12	Total
First year administered	1980	1982	1984	1987	
Content areas tested					
English-language arts			x*	x*	
Reading	x	x	x	x	
Written expression	x	x			
Direct writing assessment			1987	1988	
Mathematics	x	x	x	x	
History-social science			1985		
Science			1986		
Number of forms	30	40	60	24	154
Items per form	34	31	36	26**	
Total items	1,020	1,240	2,160	606	5,026
Items per form by content					
Reading	8	10	6	10	
Written expression	14	9	4(editing)	4(editing)	
Mathematics	12	12	7	10	
History-social science			10		
Science			9		
Number of skill scores					
Reading	27	54	13	13	107
Written expression	34	39			73
Direct writing assessment			33	17	50
Mathematics	29	50	40	9	128
History-social science			41		41
Science			40		40
Total skill scores	90	143	167	39	439
Supplementary information					
Sex	x	x	x	x	
Mobility	x	x	x	x	
English-language fluency	x	x	x	x	
Other language spoken	x	x	x	x	
SES—Parent occupation	x	x			
SES—Parent education			x	x	
Special program particip.	x	x	x		
Time reading		x	x	x	

continued on next page

Table 6–1 — *Continued*

	Grade 3	Grade 6	Grade 8	Grade 12	Total
First year administered	1980	1982	1984	1987	
Supplementary Info. Con't.					
Time watching TV		x	x	x	
Time on homework		x	x	x	
Writing assignments		x	x	x	
Attitude toward subjects	x	x			
Ethnic background	x	x	x	x	
Courses completed				x	
Extracurricular activities				x	
Grades repeated	x				
Special math questions		x			
Post high school activities				x	
School climate			x		
Open-ended mathematics				x	

NOTES
(*) At grades 8 and 12, reading and editing are combined into English-language arts
(**) 24 items per form plus two math computation items per each of 15 supplements (30 computation items total)

and difficult questions and consists of items from all major skill areas—stratified by difficulty and content. For test administration, the forms are stacked sequentially and are distributed to students in a manner similar to conventional tests. Since each student takes only one of the thirty forms containing 34 questions, testing time is limited to only one class period.

Since CAP administers tests to each student in each school, it allows for aggregating data to produce reports at the school and district levels. Although CAP's procedure could allow reports at the classroom level, no classroom reports are produced.

The report shown in Figure 6–1 is the skill area report for a typical school, showing the total score along with the subscores useful for program diagnostic purposes. The subscores are shown with a band of 0.67 standard error of measurement around the point estimate to discourage overinterpretation of skill area scores. In general, if the skill area band is below the total score line, it reflects an area of relative weakness; similarly, if the band is clearly above the total score line, it reflects an area of relative strength. If the band overlaps the total score, it is neither an area of relative weakness nor relative strength. The interpretation and meaning of these data must be judged

Figure 6–1.

CAP Survey of Basic Skills: Program Diagnostic Display for Mathematics, Grade 3—1988.

Survey of Basic Skills: Grade 3 — 1988
PROGRAM DIAGNOSTIC DISPLAY
MATHEMATICS

California Assessment Program

School: SHAMROCK ELEMENTARY
District: CALWEST UNIFIED
County: CALIFORNIA COUNTY
CDS: 99-12345-6789012

The questions on the *Survey* and the reporting of scores reflect a central concern of the *Mathematics Framework* that problem solving/applications serve as an umbrella for all mathematics strands. As shown below, the scores in all skill areas are broken down into skills and applications components. The "Applications" score under Problem Solving is an aggregation of scores for all application categories.

Interpretive Example
Your total Mathematics score of 282 is expressed below as a bar. Your score in Counting and Place Value is identified as a relative strength nor a weakness because the bar overlaps the vertical line. Each skill area score is displayed as a bold vertical line, and a relative strength nor a weakness because the bar overlaps the vertical line.

See Part IV for an illustrative description of the Mathematics skill areas tested.

Your total Mathematics score of **282** is represented by the bold vertical line.

MATHEMATICS SKILL AREAS	Scaled Score and Standard Error	Relative Strength/Weakness		
		85–86	86–87	87–88
MATHEMATICS	282 ±12			
Counting and Place Value	265 ±21			
Skills	263 ±26			
Applications	269 ±35		RW	
Operations	292 ±16			
Basic facts	306 ±43			
Addition	279 ±35	RS	RS	
Subtraction	290 ±27			
Multiplication	346 ±33			
Applications	253 ±21	RW	RW	RS
Basic facts	286 ±37			RW
Addition/subtraction	229 ±34	RW	RW	RW
Multiplication	248 ±34			
Nature of Numbers and Properties	265 ±22			
Properties and relationships	285 ±38			
Money and fractions	275 ±36			
Applications	235 ±35		RW	RW
Geometry	254 ±26			
Skills	256 ±34		RW	RW
Applications	252 ±38	RW		
Measurement	267 ±21		RS	RW
Linear measures	263 ±32			RW
Other measures	273 ±37		RS	RW
Applications	264 ±39			
Patterns and Graphs	290 ±34		RS	RS
Skills	250 ±39		RS	
Applications	330 ±54			
Problem Solving	278 ±16		RW	
Analysis and models	367 ±53			RS
Applications	265 ±16			RW

Scale: 100 150 200 250 300 350 400

professionally by curriculum experts before any curricular changes are made. The experts will take into consideration the importance of certain skill areas, their interrelationship with other skill areas, and the nature and relevance of the questions on the test.

A BRIEF HISTORY OF THE CAP

As discussed earlier, the purpose of the CAP is to provide programmatic information to schools, districts, and the state as a whole. The changes in the assessment program, therefore, should be seen in light of this purpose.

Until 1972, achievement testing in California consisted of testing students in a variety of grades with one or more commercially published standardized tests. For example, in 1972, the CAP administered the *Comprehensive Tests of Basic Skills* (CTBS), Form S at Grade 3; the CTBS, Form Q at Grade 6; and *Iowa Tests of Educational Development* (ITED) Form X, at Grade 12. However, it was soon realized that the standardized tests did not match California's curriculum. A statewide task force was established to examine the content of these tests and to make recommendations to the legislature. On the basis of the recommendations of this task force, the California legislature mandated that the CAP develop tests that would be appropriate to assess the variety of curricular programs in California.

Beginning in 1972, the California Assessment Program initiated the development of new tests at Grades 3, 6, and 12. With the help of statewide committees, CAP developed test content specifications at each grade level. For the sake of efficiency, CAP chose to lease questions from test publishers' item pools that matched the specifications, rather then writing its own questions. Since only reading was assessed at Grade 3, and reading, written expression, and mathematics were assessed at Grades 6 and 12, the use of leased publishers' items was reflected in the 1973 version of Grade 12 and the 1975 version of Grade 6 tests. Because only items from the publisher's item bank were used, the committees soon realized that the tests were limiting in scope, as reflected in the test content specifications, because items reflecting the quality of California's curriculum were not available for many of the strands of mathematics.

In 1975, the CAP started developing its own test items with the help of statewide content area advisory committees consisting of educators throughout California. The CAP constructed its first instrument in mathematics at the third-grade level in 1975, followed by the revision of the sixth-grade test in 1982, a new test at Grade 8 in 1984, and a revision of the twelfth-grade test in 1987. The following sections describe how the specifications of test content and, therefore, the nature of questions on the test, have been changed from 1975 to 1987.

THE CAP'S TEST CONSTRUCTION PROCEDURES

The California Assessment Program began constructing its own test questions in 1975. Although from a mechanical aspect test construction procedures from 1975 to date have remained the same, significant substantive changes have taken place in the nature of questions since that time. The paragraphs below first describe the mechanics of test construction followed by the changes in the specification of test questions and the writing of those questions.

Mechanics of Test Construction

Following are the main steps in the CAP's test development process for mathematics:

1. *Establishing an assessment advisory committee.* An assessment advisory committee is established consisting of curriculum specialists from the following groups: school districts, offices of county superintendents of schools, professional associations, the California State University, the University of California, and the state department of education.
2. *Reviewing existing curricular and instructional materials.* The CAP staff reviews the California *Mathematics Framework*, the state-adopted textbooks, county courses of study, and other curriculum materials, such as the *Model Curriculum Guide*, to prepare preliminary test content specifications. The members of the advisory committee review the specifications and help CAP staff write illustrative test questions.

3. *Establishing a test development team.* In addition to the assessment advisory committee, an ad hoc item-writing team, consisting primarily of classroom teachers from the appropriate grade levels, is established. The teachers are selected from a pool established from the recommendations of the advisory committee members, directors of the California Mathematics Projects, officers of the California Mathematics Council, the pool of applicants for the president's award in mathematics, and other mathematics educators having a stake in assessment. Approximately ten to fifteen teachers per grade level serve on the item-writing team.

4. *Writing test questions.* The item-writing team members are given the task of writing items. Some members of the advisory committee who have a special interest in a specific grade level also participate in the item-writing process.

Questions are written by the team members individually or jointly in small groups. In certain hard-to-measure concepts or problem-solving tasks, three or four members of the team may be engaged in discussion with perhaps two members of the team listening to the discussion, writing, and verifying with members discussing the concept that their item was the one under discussion.

For example, in 1980 when the sixth-grade test was being revised, the discussion group felt that students performed quite well on questions related to mathematical operations, but they did not understand what the different steps in the operation meant. The group wanted to provide a question in which the student did no computation but could interpret the results of a correctly performed calculation. After several trials, the item writer wrote the following question:

> 130 students from Marie Curie School want to go to a school picnic. A school bus can carry 50 students. John did the following calculation to find the number of buses needed for the picnic.

$$
\begin{array}{r}
2 \\
50 \overline{)\ 130} \\
100 \\
\hline
30
\end{array}
$$

John's arithmetic is correct. How many buses will be needed to carry all the students?

(A) 30   (B) 3   (C) 2   (D) 2R30

Of course, the above question is the final product after several reviews and edits. The point is that an item like this requires collective thinking, checking, and validation before it takes final shape.

5. *Reviewing and editing.* The item-writing team members usually meet six to eight times for two or three days each over a period of nine to twelve months. After the team members have completed the writing process, the advisory committee and the item-writing team jointly review the items. The questions are edited for clarity, appropriateness of response choices, and mathematics assessed by the items. After the committee review, the CAP staff reviews each item for consistency in format, correctness of artwork, and precision of technical writing.

6. *Field testing items.* Usually the number of questions for field testing is quite large. For example, during the test construction phase of the eighth-grade test between 1982 and 1984, approximately fifteen hundred questions were field tested in mathematics. For field-testing purposes, the questions are distributed into short forms, each form consisting of approximately thirty-five items so as to be easily administered in one class period. Each form is balanced for content and difficulty so that the student sees each form as a complete test in mathematics. All California school districts are sent an invitation to participate in field testing. School districts are also asked if their teachers would be willing to participate in item review. In this process, teachers review the questions from two test forms for clarity and indicate the degree of instructional emphasis and appropriateness of these items as a measure of the effectiveness of their district's mathematics program. Of the approximately eight hundred school districts having an eighth-grade, five hundred volunteered to participate in field testing. Approximately six hundred teachers reviewed the questions and approximately twenty thousand students participated in the field testing process.

7. *Calculating item statistics and compiling field review data.* Numerous item statistics, such as item difficulty for

each item in the group as whole, are calculated. Item statistics are also arranged by subgroup of students, such as by sex, ethnic group, language fluency group, and socioeconomic category. Item correlation with the total test is also calculated for each group and for each response choice. Several bias indices, indicating the discrepancy between the performance of a particular group and the total test population, are also calculated.

8. *Reviewing field-tested items.* Advisory committee members review the difficulty of each item and look for problems such as bias, unclear wording, inappropriate response choices, or inconsistent formats among the items to assure that only the best items survive the analysis of the field test. The items are modified or deleted based on an indication of bias and inappropriate or misleading wordings. The committee uses field test data to improve the overall quality of items. The modified items are field tested again to check whether the modifications have introduced additional unforeseen defects.

9. *Selecting the final set of items.* The advisory committee members, working with the CAP staff, select the final set of test questions. The selected questions reflect the proportions of items according to an agreed-upon distribution of items as specified in the test content specifications. For example, the distribution of items according to various reporting categories of mathematics for the sixth-, eighth-, and twelfth-grades (1984 version) is shown in Tables 6–2, 6–3, and 6–4 respectively.

Table 6–2
Skill Areas Assessed in Mathematics—Grade 6

I. **Counting, Numeration, and Place Value**
   A. Skills
      1. Counting and numeration
      2. Place value
   B. Applications

II. **Nature of Numbers and Properties**
   A. Skills
      1. Ordering and properties
      2. Classification of numbers
   B. Applications

III. Operations
A. Skills
1. Addition/subtraction of whole numbers
2. Multiplication of whole numbers
3. Division of whole numbers
4. Addition/subtraction of decimals
5. Multiplication/division of decimals
6. Operations on fractions
7. Percents and equivalent fractions and decimals
B. Applications
1. One-step involving whole numbers
2. One-step involving rational numbers
3. Two (or more) steps

IV. Expressions, Equations, and Coordinate Graphs
A. Skills
1. Expressions and equations
2. Graphs and function tables
B. Applications

V. Geometry
A. Skills
1. Shapes and terminology
2. Relationships
B. Applications

VI. Measurement
A. Skills
1. Metric units
2. U.S. Customary units
3. Perimeter, area, and volume
B. Applications

VII. Probability and Statistics
A. Probability
B. Statistics

VIII. Tables, Graphs, and Integrated Applications
A. Tables and graphs
B. Integrated applications

IX. Problem Solving
A. Formulation
B. Analysis and strategy
C. Interpretation
D. Solution of problems

Table 6–3
Skill Areas Assessed in Mathematics—Grade 8
(Total number of questions: 468)

		Percent
I. Numbers		**15**
A. Skills/concepts		10
1. Order relations and classification	3	
2. Number theory	4	
3. Properties	3	
B. Applications		5

continued on next page

Table 6–3 — *Continued*

		Percent
II. Operations		**15**
A. Skills/concepts		7
1. Whole and rational numbers	4	
2. Percents, proportions, and conversions	3	
B. Applications		8
1. One-step	4	
2. Two or more steps	4	
III. Algebra		**15**
A. Skills/concepts		10
1. Expressions and equations	5	
2. Graphs and functions	5	
B. Applications		5
IV. Geometry		**15**
A. Skills/concepts		10
1. Geometric terms and figures	4	
2. Geometric relationships and postulates	6	
B. Applications		5
V. Measurement		**9**
A. Skills/concepts		6
1. Units and estimations	3	
2. Measurement of perimeter, area, and volume	3	
B. Applications		3
VI. Probability and Statistics		**8**
A. Probability		4
B. Statistics		4
VII. Tables, Graphs and Integrated Applications		**7**
A. Tables and graphs		4
B. Integrated applications		3
VIII. Problem Solving		**16**
A. Formulation of a problem		4
B. Analysis of a problem		4
C. Strategies		5
D. Interpretation		3

Table 6–4
Reporting Categories
Survey of Academic Skills: Grade 12
Mathematics

I. **Problem Solving/Reasoning [25%]**
   A. Problem formulation
   B. Analysis and strategies
   C. Interpretation of solutions
   D. Nonroutine problems/synthesis of routine applications

II. **Understandings and Applications [75%]**
   A. Numbers and Operations [14%]
      1. Nature of real numbers
      2. Selection and use of operations on real numbers

B. Patterns, Functions, and Algebra [17%]
   1. Patterns
   2. Relations, functions, and graphs
   3. Algebra
C. Data Organization and Interpretation [18%]
   1. Organizing data as graphs and charts
   2. Statistics
   3. Probability and systematic counting
D. Measurement, Geometry, and Spatial Relationships [18%]
   1. Mensuration
   2. Geometric and spatial relationships
E. Logical Reasoning [8%]
   1. Quantifiers, Connectives, and Relationships
   2. Using deductive and inductive reasoning

10. *Reviewing the selected questions.* The final set of questions is then subjected to another review by CAP staff and testing professionals. In addition, a variety of item statistics are examined in the search for otherwise undetected defects and sources of bias. The questions are also reviewed by experts for linguistic, ethnic, and gender bias.

Test Content Specifications

Test content specifications are the blueprint for test item construction. The test content specifications denote the depth and breadth of what is considered important for assessment. They are also the bridge between curriculum/instruction on the one hand and assessment on the other. In other words, test content specifications serve as the main evidence to establish content validity of a test instrument.

The reader will discern that the procedures for delineating test content specifications in the CAP have gone through changes over time. These changes reflect the prevailing tension between the concerns of policy makers and the concerns of mathematics educators.

Specifications in 1975: Grade 3 Test Development. During the period in which the third-grade test was developed, the prevailing philosophy of test development was that in order for the tests to be accepted by a vast majority of districts for their program evaluation, the test must match what was actually being taught in their classrooms. Furthermore, since what was being taught in the classrooms was based on state-adopted textbooks, the content of the test had to be limited to what

appeared in most textbooks. Therefore, the test content specifications were written based upon the content appearing in state-
adopted textbooks at the time.

Figures 6–2a, 6–2b, and 6–3a, 6–3b show the pages appearing in the draft Test Content Specifications-Operations for the
Third-Grade Test (1980 version). In Figures 6–2a and 6–3a, one
column provides the page numbers of the textbook containing
a particular topic. Figures 6–2a, 6–2b and 6–3a, 6–3b show, in
particular, that certain mathematical content, such as basic
arithmetic, appeared in all textbooks; however, topics such as
problem solving and modeling did not appear in any of the
books, or appeared in only one or two books. Before developing
the final test content specifications, a random sample of teachers from throughout the state was surveyed to determine the
degree of emphasis they placed on each skill and whether they
would like that skill to be measured as part of the statewide
assessment. The resulting specifications were quite narrow in
the sense that important mathematical topics, such as problem
solving and modeling, were not taught in most classrooms.

Specifications in 1980: Grade 6 Test Development. The
test content specifications for the sixth-grade test, developed
between 1978 and 1980, were derived from the *Mathematics
Framework for California Public Schools* rather than exclusively
from the content analysis of commonly used state-adopted textbooks. The *Agenda for Action*, published by the National Council of Teachers of Mathematics, was also influential in developing the test content specifications. As a result, it was determined
appropriate to include a problem-solving subtest, such as problem formulation, problem analysis, and problem interpretation,
in addition to an emphasis on routine and nonroutine problem-
solving skills. The specifications also included skills in geometry, algebra, measurement, and probability and statistics. Table
6–2 above shows the content outline of the sixth-grade test.

Specifications in 1984: Grade 8 Test Development. The
test content specifications for the eighth-grade test, developed
between 1981 and 1984, were based on the *Mathematics Framework for California Public Schools* and the *Model Curriculum
Guide*. The rationale for the test content specifications was
based on three major concerns: (1) a general concern for "excellence" in that all children deserve a decent education involving

Figure 6–2a.
CAP Test Content Specifications—Operations for the Third–Grade Test (1980 version).

Skill	Performance Objective	Item Stem Characteristics	Response Characteristics	Text Book Page Numbers	
Recall basic addition facts	Given two numbers, the student will add them and select the correct answer from four options.	The item stem will be two single digit numbers aligned either vertically or horizontally with a proper addition sign. Note: A list of basic addition facts, categorized by level of difficulty is given in the appendix.	The response choices will be whole numbers with one and two digits. The incorrect options will be the result of: a. making an error in the basic facts (one more or one less than the correct answer) b. perceptual error of reversing the sum digits (e.g., 8 + 9 = 71) c. failing to understand the meaning of the operation sign (e.g., merging the addends; 7 + 3 = 73) d. failing to interpret the operation sign correctly (e.g., multiplying instead of adding) e. failing to understand the identity element zero (when it occurs in the stem); e.g., choosing zero as the correct answer	SF Ho HM He AW SRA SB	14–23, 24–31, 32–33 46–49, 62–69, 74, 79, 101, 110 2–4, 36–46 1, 7, 8, 11–12 25–26, 35–36, 46, 55, 70–71, 73–78, 83–84 28–35, 40–41, 50–53, 59–61, 64, 87 6–9, 14, 60, 152–154, 287 28–30, 36, 39, 43–50, 52 58–60, 328

Figure 6-2a — Continued

Skill	Performance Objective	Item Stem Characteristics	Response Characteristics	Text Book Page Numbers
Recall basic subtraction facts	Given two numbers, the student will find the difference and select the correct answer from four options.	The item stem will be a one- or two-digit minuend and a single-digit subtrahend aligned vertically or horizontally with a properly positioned subtraction sign. Note: An extensive list of subtraction facts is given in the appendix.	The response choices will be whole numbers with one or two digits. The incorrect options will be the result of: a. making an error in the basic facts (one more or one less than the correct answer) b. failing to understand the meaning of the operation sign (e.g., moving the minuend and subtrahend together) c. failing to interpret the operation sign correctly d. failing to understand the identity element zero (when it occurs in the stem); e.g., choosing zero as the correct answer e. reversal of direction of subtraction—subtracting top number from bottom number	SF 14–23, 24–31, 32–33 Ho 50–52, 55, 64 67–69, 74, 101, 107, 110 HM 5–8, 10, 29–30, 37–38 He 3–4, 6, 13–17, 25–26, 35–36, 46, 55, 70–71, 73–78, 83–84 AW 28–35, 40–41, 50–53, 59–61, 64, 123 SRA 13–16, 152–154, 289 SB 33–39, 43–50, 52, 58–60, 329

SF: Scott, Foresman and Company—Mathematics Around Us: Skills and Applications, by Bolster, Level 3, Pupil's Book, 1975
Ho: Holt Rinehart and Winston—Holt School Mathematics, Nichols, et al., Student Text, 1974
HM: Houghton Mifflin Company—Mathematics for Individual Achievement, by Denholm, Level 3 text, 1974
He: D.C. Heath and Company—Heath Mathematics by Dilley, et al., Level 3 Student Edition, 1975
AW: Addison-Wesley Publishing Company, Inc.—Investigating School Mathematics, by Eleenor, R., Eicholz, P., O'Daffer, Book 3, 1976
SRA: Science Research Associates—Mathematics Learning System, Level 3, 1974
SB: Silver Burdett Company—Silver Burdett Mathematics System, by LeBlanc, Level 3, Student Edition, 1973

Figure 6–2b.
Illustrative Examples of Operations Items from the CAP Third-Grade Test Content Specifications (1980 version).

Example 1	Example 2	Example 3	Example 4	Example 5
$\begin{array}{r} 2 \\ +2 \\ \hline \end{array}$	$6 + 0 = \square$	$\begin{array}{r} 6 \\ +1 \\ \hline \end{array}$	$8 + 1 = \square$	$\begin{array}{r} 8 \\ +4 \\ \hline \end{array}$
A. 22 (c) B. 3 (a) C. 4 (*) D. 5 (a)	A. 6 (*) B. 0 (e) C. 60 (c) D. 7 (a)	A. 61 (c) B. 7 (*) C. 5 (d) D. 8 (a)	A. 10 (a) B. 7 (d) C. 9 (*) D. 81 (c)	A. 84 (c) B. 4 (d) C. 21 (b) D. 12 (*)
$3 - 0 = \square$	$\begin{array}{r} 6 \\ -1 \\ \hline \end{array}$	$10 - 6 = \square$	$\begin{array}{r} 15 \\ -7 \\ \hline \end{array}$	$12 - 9 = \square$
A. 0 (d) B. 2 (a) C. 3 (*) D. 30 (b)	A. 5 (*) B. 7 (c) C. 4 (a) D. 61 (b)	A. 4 (*) B. 5 (a) C. 16 (c) D. 106 (b)	A. 8 (*) B. 22 (c) C. 12 (a) D. 22 (c)	A. 3 (*) B. 4 (a) C. 17 (e) D. 129 (b)

Figure 6-3a.
CAP Test Content Specifications—Problem Solving and Logical Thinking for the Third Grade Test (1980 version).

Skill	Performance Objective	Item Stem Characteristics	Response Characteristics	Text Book Page Numbers
Identify an appropriate question or problem related to a number sentence	Given a number sentence or geometric model, the student will determine the appropriate question or story problem that is related to it and select the correct answer from four options.	The item stem will be a simple whole number sentence or simple geometric shape, with directions to choose a matching story problem or an appropriate question.	The response choices will be simple story problems or simple questions. Incorrect options will be the result of: a. failing to choose correct operation b. failing to choose correct numbers c. choosing incomplete information d. choosing inappropriate question e. failing to choose correct shape f. choosing measurement question	SF — Ho 56–57 HM — He — AW — SRA — SB —
Identify an appropriate analysis for a story problem	A. Given a story problem and question, the student will determine the correct procedure to solve the problem and select the correct answer from three or four options.	The item stem will be a simple story problem with directions to choose the appropriate operation, table, graph, diagram, or estimated answer.	The response choices will be simple graphs, diagrams, tables, or the written words *add, subtract, multiply, divide*. The incorrect options will be the result of: a. failing to choose correct operation b. failing to associate the correct graph, table, or diagram with the given information	SF 35, 145, 293 Ho 157 HM B28 He — AW — SRA — SB —

SF: Scott, Foresman and Company—Mathematics Around Us: Skills and Applications, by Bolster, Level 3, Pupil's Book, 1975
Ho: Holt Rinehart and Winston—Holt School Mathematics, Nichols, et al., Student Text, 1974
HM: Houghton Mifflin Company—Mathematics for Individual Achievement, by Denholm, Level 3 text, 1974
He: D.C. Heath and Company—Heath Mathematics by Dilley, et al., Level 3, Student Edition, 1975
AW: Addison-Wesley Publishing Company, Inc.—Investigating School Mathematics, by Eleenor, R., Eicholz, P., O'Daffer, Book 3, 1976
SRA: Science Research Associates—Mathematics Learning System, Level 3, 1974
SB: Silver Burdett Company—Silver Burdett Mathematics System, by LeBlanc, Level 3, Student Edition, 1973

Figure 6–3b.
Illustrative Examples of Problem Solving and Logical Thinking Items from the CAP Third-Grade Test Content Specifications (1980 version).

Example 1	Example 2	Example 3	Example 4	Example 5
$5 - 3 = 2$	$10 + 6 = 16$	$6 \times 2 = 12$	Which question would you ask?	Which question would you ask?
Which matches the problem?	Which matches the problem?	Which matches the problem?		
A. 3 apples, 5 apples. How many in all? (a)	A. Rico had 10 marbles. He gave 6 away. How many were left? (a)	A. 3 cars, 2 people in each car. How many people? (b)	A. How big is the triangle?	A. How many triangles do you see? (*)
B. 3 apples, 5 seeds. How many seeds? (a)	B. Rico had 10 marbles. He got 6 more. How many in all? (*)	B. 6 cars, 2 more cars. How many cars in all? (a)	B. How many were left? (d)	B. What is the sum? (d)
C. Jan had 5¢. She spent 3¢. How much was left? (*)	C. Sam had 6 marbles. He lost 4. How many were left? (a)	C. 6 cars, 2 drove away. How many were left? (a)	C. How many circles? (e)	C. How much left? (d)
D. Jan had 3¢. She spent 2¢. How much was left? (b)	D. Sam had 6 marbles. Each had 4 dots. How many dots in all? (b)	D. 6 cars, 2 people in each. How many people in all? (*)	D. How many sides are there? (*)	D. How far around the square (e)

continued on next page

Figure 6-3b — *Continued*

Example 1	Example 2	Example 3	Example 4	Example 5
Ann had 4 apples. Joe gave her 2 more. How many did she have all together?	Alice spent 15¢ for pencils. She bought 5 pencils. How much for each pencil?	30 people 10 like vanilla ice cream. 5 like strawberry ice cream. 15 like chocolate ice cream.	There are 12 people. 6 small 4 middle 2 big	There were 2 boys. Each boy had 3 cookies. How many cookies in all?
How would you find the answer?	How would you find the answer?	Which shows this?	Which shows this?	Which shows this?
A. Subtract (a) B. Add (*) C. Multiply (a) D. Divide (a)	A. Divide (*) B. Subtract (a) C. Multiply or Add (a)			

higher-level thinking, problem solving, and understanding; (2) a commitment to research that expands knowledge and understanding of how students develop thinking skills and learn to solve problems; and (3) a concern that the quality of test questions was less than desirable on standardized tests.

The theme of higher expectations and improved achievement is addressed in the *Mathematics Framework for California Public Schools (1985)*, which states:

> The mathematics program recommended in this framework reflects raised expectations for student achievement. The goal for all students is to be able to use mathematics with confidence; therefore, every student must be instructed in the fundamental concepts of each strand of mathematics and no student limited to the computational aspects of the number strand. . . . Most students will go beyond the fundamental concepts to achieve deeper and broader capability in mathematics, but even the less capable students, by learning these concepts, will have appropriate experiences in all of the strands. They must not, for example, be deprived of work in geometry or probability in order to have more practice with narrow computational skills. Rather, they will continue to learn the new concepts of all of the strands and to integrate those concepts into their understanding throughout their school careers. . . . This expectation applies to all students, including students with special needs and those who come from groups who have historically been underrepresented in upper level mathematics courses. (p. 3)

In addition to the concern for excellence, a major emphasis of the *Mathematics Framework for California Public Schools: Kindergarten Through Grade Twelve* (1985) and the *Mathematics Model Curriculum Guide* (1987) is teaching for understanding. The theme of teaching for understanding is stated in the *Mathematics Framework for California Public Schools* (1985):

> Teaching for understanding emphasizes the relationships among mathematical skills and concepts and leads students to approach mathematics with a common-sense attitude, understanding not only how but also why skills

are applied. Mathematical rules, formulas, and procedures are not powerful tools in isolation, and students
who are taught them out of any context are burdened
by a growing list of separate items that have narrow
application. Students who are taught to understand the
structure and logic of mathematics have more flexibility
and are able to recall, adapt, or even recreate rules
because they see the larger pattern. Finally, these students can apply rules, formulas, and procedures to solve
problems, a major goal of this framework. (p. 12)

The concern for excellence and an emphasis on understanding in the *Framework* resulted in a consensus that the
eighth-grade test must have the following characteristics.

- In computational questions, emphasis was placed on
  the understanding of an arithmetic operation rather
  than on performing an algorithmic manipulation. Most
  of the questions can be answered mentally if the student has a clear understanding of arithmetic operations and symbols.
- Test questions reflected a level of achievement and
  sophistication consistent with a mathematics program
  that eliminates the repetition of content from one grade
  level to the next unless there is an increase in depth
  or breadth.
- Test questions were designed to assess not only the
  arithmetic computational skills, but also the skills involved in pre-algebra, geometry, measurement, logic,
  and probability and statistics.
- Special emphasis was placed on the assessment of
  problem-solving processes, such as problem formulation, problem analysis, interpretation of results, and
  on problem-solving questions, including routine and
  nonroutine problems.

The art of writing questions in problem solving and other
hard-to-measure areas was influenced by the work of Lester
(1978, 1982), Lesh (1983), Mayer (1983), Newell and Simon
(1972), Polya (1957, 1965), Resnick (1983), Schoenfeld (1982),
Sternberg (1981, 1983), and Silver (1982). Pandey (1983) described the implications of research in problem solving for

assessment. The California Assessment Program's efforts to improve the quality of test questions were aided by the work of the MAAC members and through the review of test items by Alan Hoffer, University of Oregon; Thomas Romberg, University of Wisconsin; James Wilson, University of Georgia; and Mary Kay Corbitt, East Tennessee State University.

Specifications in 1987: Grade 12 Test Development. The test content specifications for the twelfth-grade test, developed between 1983 and 1985, was based on the desire to raise expectations for all students and to develop mathematical power in students before they graduate from high school. The *Framework* (1985) defines mathematical power as follows:

> To enable all graduates to meet current and future demands, mathematics education must focus on students' capacity to make use of what they have learned in all settings. Mathematical power, which involves the ability to discern mathematical relationships, reason logically, and use mathematical techniques effectively, must be the central concern of mathematics education and must be the context in which skills are developed. (p. 1)

The major difference between the revised twelfth-grade CAP test—now called Survey of Academic Skills, Grade 12, and the older version called the Survey of Basic Skills, Grade 12, is that the new test emphasizes understanding of mathematical concepts and problem solving—a shift in emphasis similar to that of the eighth-grade test. The specifications were written and test questions were designed to measure what students understand about the mathematical concepts and skills they have learned from kindergarten through Grade 12 and how well they can use this learned mathematics in familiar and unfamiliar problem situations. The test was designed to assess students' abilities to estimate, to discern relationships, and to use number sense in the evaluation and interpretation of intermediate and final results of a problem-solving process. It requires students to use higher-level thinking skills and therefore provides a measure of their ability to do so in a mathematical setting as opposed to providing a measure only of their ability to perform rote mathematical algorithms which they may do correctly but do not understand. Table 4 shows the skill areas reported for the twelfth-grade test. In comparing the skill areas reported for Grade 12 with the skill areas reported for the sixth- or eighth-grade test (Tables 6–2 and 6–3), it is interesting to note that problem solving was the skill area reported first, followed by another major skill area, understanding.

The Art of Questioning

To appreciate how the CAP has evolved over a period of years, we will examine relationships among test specifications, the nature of the questions, and the reporting categories over a period of fifteen years. In the CAP's evolution, it should be emphasized that CAP test designers have consistently sought to blend into their assessment instruments the most current knowledge about our understanding of the nature of mathematics, theories of learning, art of test construction, and program improvement strategies.

Early Developments

Third-grade test development in 1975 was heavily influenced by the writings of Popham (1973) and Millman (1974), who recommended rigorous specifications to ensure that each item be a true reflection of the intended skill to be measured. The structure of the test content specifications was derived using the traditional content-by-process matrix. For example, the content categories for the third-grade specifications were the strands of mathematics specified in the California *Mathematics Framework for Kindergarten through Grade Twelve* (1985)—Number, Algebra, Geometry, Measurement, Probability and Statistics, and Logic. The process categories were computation/knowledge of facts, comprehension, and applications. Detailed specifications were generated for each cell of the content-by-process matrix. As shown in Figures 6–2a and 6–2b, the item specifications were very structured in the specificity of the performance mode and characteristics of distractors. The performance mode described the limits of the item stem of a multiple-choice question, and distractor characteristics described the various ways of constructing the incorrect choices for the item.

This method of test construction resulted in a large number of items, each item designed to measure a discrete skill described in the specification. The collection of items in a sub-domain, such as functions, contributed to the sub-domain scores.

This analytical approach toward test construction is based on the assumption that a single content area, such as mathematics, can be divided into measurable minuscule bits and pieces, that a sampling of behaviors from the content-by-process matrix could be generalized to the population of behaviors, and that the diagnostic information generated from sub-domain scores will be useful for improving a school's program.

This method of constructing achievement tests has long proved useful. It is still useful in situations where instruction is not "corrupted" by *how* tests are constructed, or where test validity is not tampered with by revealing the exact content of the test. However, in situations of high stakes testing, as in situations where teachers or administrators are held accountable for student achievement, the analytical approach is undesirable. When the test questions can be replicated through precise test specifications and the results reported for a large number of narrowly defined skill areas, the message is conveyed that learning can be improved by teaching bits and pieces of information. In many situations, rather than teaching to desirable curriculum practices, teachers resort to "multiple-choice" instruction because of the form of the assessment instrument.

Recent Developments

In recent years, as in the development of the twelfth-grade test implemented in 1985, criteria for item specifications take into account the emerging role of high stake tests. The criteria can be traced to three main concerns:

1. Test questions must reflect the current view of the nature of mathematics. This view emphasizes understanding, thinking, and problem solving that require students to see mathematical connections in a situation-based problem and to be able to monitor their own thinking processes to accomplish the task efficiently. This requires that test questions have the following characteristics:

    • They assess thinking, understanding, and problem solving in a situational setting as opposed to algorithmic manipulation and recall of facts.
    • They assess the interconnection among mathematical concepts and the outside world.

2. Test questions must reflect the current understanding of how children learn. The current view
of instruction and learning assumes that children
are active learners and engage in creating their
own meaning during the instructional process. This
requires that test questions have the following
characteristics:

- They must be engaging.
- They must be situational and based upon real-
life applications.
- They must have multiple-entry points in the sense
that students at various levels in their mathematical sophistication should be able to answer
the question.
- They allow students to explore difficult problems
and students' explorations are rewarded.
- They allow students to answer correctly in diverse ways according to their experiences, rather
than requiring a single answer.

3. Test questions must support good classroom instruction and not lend themselves to distortion of
curriculum. Good curriculum practices require that
test questions have the following characteristics:

- They must be exemplars of good instructional
practices.
- They should be able to reveal what students know
and how they can be helped to learn more
mathematics.

Questions having these characteristics have been christened
by Honig (1985) as "power items." Such questions cannot be
measured by the typical test comprising thirty- to sixty-second
multiple-choice questions. However, multiple-choice questions
requiring two to four minutes can be developed that have most
of the characteristics described above. Examples 1 and 2 (see
Appendix D) are questions of the type appearing in the twelfth-
grade test, the Survey of Academic Skills: Grade 12. In the
twelfth-grade test, CAP also uses open-ended questions that
require 12 to 15 minutes for students to answer. Example 3
has been taken from the 1987–88 version of the test, which has
been discussed in detail in *A Question of Thinking* (California

State Department of Education, 1989). Example 4, shown in Appendix D, also shows the response of one student on this question.

CAP Instruments in the Future. This paper describes CAP test development procedures prior to 1989. The CAP is currently revising tests at Grades 3, 6, 8, and 12 and introducing a new test at Grade 10. Besides the new type of multiple-choice questions, the revised tests will have open-ended and performance-type questions. In addition, portfolio assessment in mathematics is being explored as an alternative. Sample performance tasks and guidelines for the portfolio can be obtained by writing to the author.

7

Assessing Students' Learning in Courses Using Graphics Tools: A Preliminary Research Agenda

Sharon L. Senk

Recently mathematics educators have called for the use of calculator and computer graphing technology in mathematics classes, and several software and curriculum development projects have attempted to transform these recommendations into reality. However, until now there has been little systematic study of how teaching, learning, and assessment in courses using such graphics tools are affected by the technology. This paper describes a preliminary agenda developed by researchers in the field for assessing students' learning in courses using graphics tools. Included are suggested investigations of student and teacher outcomes and a discussion of methodological issues.

In recent years there have been many calls for the reform of mathematics education in the United States. Among the most consistent recommendations is that mathematics programs take advantage of the power of calculators and computers (College Board, 1985; Fey, 1984; NCTM, 1980, 1989). Specifically, function graphing tools available on calculators and computers are suggested as a means to produce both a richer mathematics curriculum and a deeper understanding of mathematics without having to use valuable time in mathematics classes for a study of computer programming (Demana & Waits, 1990; Fey, 1989; Kaput, 1989; Waits & Demana, 1989).

These calls for using graphing technology have been accompanied by several research and development projects. The Educational Technology Center has developed software called *Visualizing Algebra: The Function Analyzer* (Harvey, Schwartz, & Yerushalmy, 1988) and used it to study students' conceptions and misconceptions of scale in graphs (Goldenberg, 1988). Fey and Heid (1987) are developing booklets for students, guides for teachers, and correlated computer software for elementary algebra. The University of Chicago School Mathematics Project (UCSMP) has developed a course called Functions, Statistics and Trigonometry with Computers (Rubenstein et al., 1988) in which students use as tools any standard graphing software, a statistics package, and BASIC programs. The Ohio State University Calculator and Computer Precalculus (C^2PC) Project has developed software (Waits & Demana, 1988) and designed a precalculus text (Demana & Waits, 1989) that can be used with its software or a graphics calculator at both the high school and college levels. During the 1988–89 academic year, each of the latter three groups studied the effects of their materials on learning and teaching in regular classroom settings (Demana, et al., in preparation; Lynch, Fischer, & Green, 1989; Sarther, Hedges, & Stodolsky, in preparation).

Materials based on graphing tools such as those above are reshaping the profession's conceptions of what we ought to teach, what we can teach, and how we can teach it (Fey, 1989). They allow students to explore "advanced" mathematical functions without having to master much prerequisite algebraic manipulation. Making graphs becomes a tool for solving other problems, rather than an end in itself. From exploring multiple instances, generalizations can be formed; and, conversely, instances of proposed generalizations can be tested quickly using graphing tools.

Surprisingly, there has been little discussion among mathematics educators of the methods and materials used to assess learning and teaching in such environments. Goldenberg (1988) reports that he found little in the research literature on learning or teaching about graphing functions in any environment—with or without computers.

At present there are no nationally available mathematics tests that require calculator or computer use. The Mathematical Association of America's Calculator-Based Placement Test Program Project and the College Board's Mathematics Achievement Test Committee are presently developing tests that include calculator-active items. Such tests demand changes from typical achievement tests in the types of problems that can and cannot be included (Harvey, 1989). In each case, however, the tests being developed consist only of multiple-choice items and assume that students have only a non-graphics calculator available. In addition, these tests are designed to assess students' knowledge of the present mathematics curriculum where calculator and computer use may not have been an integral part of the course. Furthermore, no guidelines exist for assessment of student learning in calculator- or computer-based courses. Recently, the NCTM (1989) has called for broadening our view of appropriate assessment techniques in all areas of mathematics. Senk (1989) and Wiske et al. (1988) have called for further research on techniques and instruments for assessing students' learning in advanced technological environments.

Call for a Meeting

Given the needs outlined above, funding was secured from the National Center for Research in Mathematical Sciences Education for a meeting to discuss ways of assessing the impact of function graphing tools on students' learning. The meeting took place on December 15–16, 1988, on the campus of the University of Chicago, with the following people participating:

> Dora Aksoy, Department of Education, The University of Chicago;
>
> James Flanders, Department of Mathematics and Statistics, Western Michigan University;
>
> E. Paul Goldenberg, Education Development Center, Newton, MA;
>
> John G. Harvey, Department of Mathematics, University of Wisconsin—Madison;
>
> M. Kathleen Heid, Department of Curriculum and Instruction, Pennsylvania State University;

Catherine Sarther, Departments of Mathematics and Education, Mount Mary College;

Sharon L. Senk, Department of Education, The University of Chicago;

Bert K. Waits, Department of Mathematics, The Ohio State University; and

Orit Zaslavsky, Department of Education in Technology and Science, Technion, Israel.

The contributions of Goldenberg, Harvey, Heid, Senk, and Waits to issues related to graphing technology are noted above. At the time of the meeting Aksoy, Flanders, and Sarther were all doctoral students at the University of Chicago working with the University of Chicago School Mathematics Project. Aksoy and Flanders were editors, and Sarther was coordinator of the 1988-89 field study of *Functions, Statistics, and Trigonometry with Computers* (Rubenstein et al., 1988). Zaslavsky was a postdoctoral researcher at the Learning Research and Development Center, University of Pittsburgh, working on a review of the literature on functions and graphs in mathematics (Leinhardt, Zaslavsky, & Stein, in press).

The two main questions this meeting addressed with respect to algebra and precalculus courses based on function graphing tools were:

1. What are the fundamental goals to be assessed? (What are the core content, processes, and beliefs?)

2. How should we go about assessing them? (What kinds of problems or situations appropriately measure these goals? What techniques enable the researcher or teacher to uncover likely causes of students' difficulties? To what extent should assessment instruments use graphing technology? To what extent should assessment be done without access to graphing tools?)

Both the funding agency and the participants hoped that the meeting would encourage the formation of an "invisible college" of researchers interested in this topic who would continue to collaborate on common interests even after the meeting was over.

Summary of Discussion

Between them the participants had used graphing technology with students at each level from Grade 9 through the college sophomore level. During the first part of the meeting each participant described briefly his or her experiences using this technology and some issues his or her own project faced related to the use of graphing tools.

Virtually everyone agreed that research or curriculum development projects could not (or at least, should not) infuse graphing technology into a secondary or college course without changing some of the original curricular goals. In particular, participants agreed that courses which use graphing technology in significant ways should, in comparison to standard courses, increase emphasis on realistic applications of mathematics; they should also focus on problems that encourage exploration and conjecturing, and decrease emphasis on many traditional manipulative skills.

The following were identified as issues faced by students and teachers in all the projects represented at the meeting:

1. mastering the technology itself (ease of use is critical for implementation)
2. balancing exact and approximate answers, coping with multiple answers
3. putting the control of instruction and learning more with students than ever before
4. worrying about long term effects of less manipulative skill and more graphical representation on students' performance in subsequent courses.

Based on the research of Fey and Heid (1987), Goldenberg (1988), and Leinhardt, Zaslavsky, and Stein (in press), scaling seems to be a large issue early in the study of functions and graphs. Beginning algebra students seem to need instruction on how changes in scale do not change the values on the graph, but only their perception of its shape or the amount of the graph they can see on the screen. Beginning algebra students in courses that use graphics tools also seem to need more explicit instruction than they do in traditional algebra courses on deciding what scale to use on graphs. However, scaling seems to be much less an issue by the time a student reaches a precalculus course in high school or college. Harvey and Waits reported that older students seem to have few difficulties changing the scale on a viewing window so a complete graph can be seen.

A Preliminary Research Agenda

The discussion then turned to the need for research on issues related to assessment in courses that use graphing technology. Participants agreed that a useful point of departure for this discussion was Standard 6 on Functions (NCTM, 1989). Our recommendations for research are grouped into two areas. Under student and teacher outcomes, we include what we believe to be the most important goals with respect to both content and process for courses which emphasize graphs and functions. Under methodological issues, we identify how we believe we should go about investigating student and teacher outcomes.

Student and Teacher Outcomes. We recommend that studies be developed to investigate the effects of graphing technology on the ability of students to:

1. interpret information from graphs alone, that is, without algebraic formulation
2. translate across representations, that is, from one tabular, graphical, function rule, or physical context to another
3. generate examples of particular types of functions, for example, linear or exponential functions
4. discuss the effects of changes in the viewing rectangle on their perception of the shape of a graph or the nature of its properties
5. solve equations or inequalities, or systems of equations or inequalities by both standard paper-and-pencil algorithms and graphical means
6. use graphs to hypothesize whether two algebraic expressions are identically equal
7. for a given function, describe its properties (behavior), for example, intercepts, maxima/minima, end behavior, points of discontinuity
8. describe the effects of parameter changes on a function within whatever representation system is used.

We also recommend that research address the impact of graphing technology on students':

9. frequency and proficiency of use of such technology

10. ability to generate higher-order questions about functions

11. ability to justify conclusions both visually, using a graphing tool, and deductively, based on properties of functions

12. beliefs about mathematics, for example, the extent to which it is fun, dynamic, or evolving

13. attitudes toward learning mathematics, such as confidence or persistence.

We further recommend that research investigate the impact of graphing technology on teachers':

14. frequency and proficiency of use of such technology

15. structure of class time (we hypothesize less one-way lecture, and more discussion and attention to students' questions)

16. beliefs about mathematics, for example, the extent to which it is fun, dynamic, or evolving

17. beliefs about learning and teaching, such as, the willingness to give messier examples, or the willingness to say, "I don't know"

18. ability to assess what students are learning, what misconceptions they have, and how students acquire their knowledge.

Methodological Issues. We believe that assessment of teaching and learning in courses using graphing tools should address the following issues.

1. *Development of instruments.* Three types of instruments are suggested for assessing students' knowledge of the content above: (a) items presented electronically, say on a computer, on which the student also has access to graphing tools; (b) items presented on paper that a student may respond to with access to graphing tools; and (c) items done completely by paper and pencil without access to graphing tools. Comparing results of studies

using all three types of items will reveal how much more information about a student's abilities a graphing utility makes available and will help determine the financial and time costs of each type of assessment instrument. Concomitant with the development of instruments that encourage use of graphing technology, we must also develop new instruments that assess the knowledge we want students to be able to apply without access to sophisticated technology. For both "button pushing" and "lead pushing" knowledge we encourage the development of both short-answer and longer, more elaborate open-ended assessment items.

2. *Variations in the technology itself.* Studies should be conducted to compare and contrast ease of use of graphing calculators and different configurations of graphing software and hardware and their effects on learning and teaching. In particular, the effects of software that allow several views of a graph simultaneously, or simultaneous views of graph and table of values, should be studied.

3. *Duration of study.* Both short-term and longitudinal studies with multiple-time-point data are suggested. The former allows researchers to get quick feedback and make revisions in curriculum or instruction based on unsatisfactory results. The latter is necessary to study cumulative effects.

4. *Nature of the research.* Both basic research (e.g., laboratory or case studies) and classroom research (e.g., curriculum evaluation) are necessary. Laboratory research using state-of-the-art hardware, software, and delivery systems, and a small number of students allows investigators to probe more deeply into what is ultimately possible for teaching, learning, and assessment. Classroom research using the best commercially available products and normal classroom conditions allows policy makers to think about what is realistic in the immediate future.

5. *Cooperative efforts.* Research on the effects of technology on learning or on methods of assessment using calculators and computers should be shared with and, on some occasions, conducted with the cooperation of professional organizations, such as the National Council

of Teachers of Mathematics or the Mathematical Association of America, and testing agencies, such as the Educational Testing Service.

Outreach and Communication

The participants suggested the following activities as appropriate next steps toward implementing the above research and development agenda:

1. Propose a symposium sharing the above ideas, and some of the results of our own investigations at the 1989 Psychology of Mathematics Education/North American Chapter Meeting PME/NA meeting.
2. Encourage dialogue with others interested in research and development activities related to graphing technology and assessment.
3. Pick some abilities on the preceding list of student outcomes, develop items measuring those abilities, and share items and results with each other.
4. Lobby for the use of calculators and computers and for technology-based tests on national, state, and local assessments, and college entrance and placement exams.
5. Define fundamental "lead pushing" skills related to graphing and functions and work to have items measuring them incorporated into standard assessment instruments.
6. Write a position paper on issues related to assessment for publication in professional journals, for instance, in the "Soundoff" section of the *Mathematics Teacher.*

As of this time (June 1990), we have accomplished Item 1 and have made some progress on Items 2 to 5. Waits and Senk shared with the other participants copies of their texts (Demana & Waits, 1989; Rubenstein et al., 1988) and selected tests used in program evaluation. (See Figures 7–1 and 7–2). Harvey and Senk organized a symposium on Changes in Student Assessment Occasioned by Function Graphing Tools at the PME/NA meeting held in September 1989, in which they, Heid, and

Waits participated. At the meeting the four decided to share items and results from their work with the hope of collecting an item bank that might eventually be used as a source for research or classroom or program evaluation. Harvey continues to work through the Mathematical Association of America on lobbying the College Board to incorporate use of technology on instruments developed by the Educational Testing Service. Waits has organized two conferences on Technology in Collegiate Mathematics, and plans to host a third in November 1990. Finally, Harvey and Senk are preparing an analysis of assessment issues related to functions and graphs for another publication.

1. Use the graph to solve $f(x) > g(x)$.

    A. $x > 0$

    B. $-2 < x < 7$

    C. $x < -2$ or $x > 7$

    D. $3 < x < 30$

    E. $x < 3$ or $x > 30$

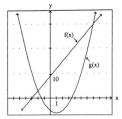

2. Which one of the following could represent a complete graph of $f(x) = x3 + ax$ where a is a real number?

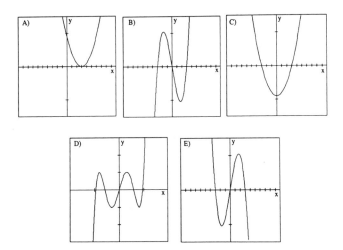

Figure 7–1. Sample multiple-choice items testing graphical knowledge of functions (Demana & Waits, 1989). Used with permission.

1. (a) Determine to the nearest tenth the zeroes of the function defined by $f(x) = x^4 - 2x^3 - 10x + 5$.

    (b) Explain your method.

2. The polynomial function A defined by $A(x) = -.0015^3 + .1058x$ gives the approximate alcohol concentration (in percent) in an average person's bloodstream x hours after drinking about 250 ml of 100-proof whiskey. The function is approximately valid for values of x between 0 and 8. How many hours after the consumption of this much alcohol would the percentage of alcohol in a person's blood be the greatest? Express the answer correct to the nearest tenth, and explain how you got your answer.

Figure 7–2. Sample open-ended items testing ability to use technology to solve problems about functions (Sarther, Hedges, & Stodolsky, in preparation). Used with permission.

8

Mathematics Testing with Calculators: Ransoming the Hostages

John G. Harvey

> We must ensure that tests measure what is of value, not just what is easy to test. If we want students to investigate, explore, and discover, assessment must not measure just mimicry mathematics. By confusing means and ends, by making testing more important than learning, present practice holds today's students hostage to yesterday's mistakes.
>
> (MSEB) Everybody Counts

This paper analyzes research on the use of calculators in mathematics testing. Three kinds of tests are considered: (a) *calculator-passive tests* (i.e., tests on which calculator use is not intended), (b) *calculator-neutral tests* (i.e., tests that have no "calculator sensitive" items and on which calculator use is not required), and (c) *calculator-based tests* that were developed so that most students will need calculators while responding to some of the items. The effects of calculator use on the characteristics of all three kinds of mathematics tests is reported. Included in the paper are examples of items from calculator-neutral tests and of calculator-active items from calculator-based tests.

The hand-held calculator was invented by Texas Instruments Incorporated (TI) in 1967. In 1972 Texas Instruments introduced the TI Data Math calculator—a four-function calculator that retailed at the time for $150. Also in 1972, the Hewlett-

Packard Company began marketing the HP–35, a scientific calculator that retailed for $395. In that era, it seemed unlikely that hand-held calculators would ever be consistently and widely used in mathematics instruction for two reasons: they were expensive, and it was unclear in what ways they could be used effectively to improve mathematics learning and teaching. The HP–35 had clearly been designed for use by engineers and engineering students; the initial market for the TI Data Math was business and industry.

In the interim ownership of hand-held calculators has become widespread. Hand-held calculators so dominate the calculator market that most people assume hand-held calculators are being discussed whenever the word *calculator* is used. Along the way calculators have become increasingly versatile; presently, the kinds of calculators range from four-function calculators with arithmetic operating systems (e.g., the TI–108) to graphics and symbolic mathematics calculators like the HP–28S. In between are nonprogrammable and programmable scientific calculators, calculators designed especially for business applications, and scientific graphing calculators that may have matrix functionality. Almost all of the calculators that have scientific functionality also have one- or two-variable statistics functionality.

The prices of calculators like those first produced by Hewlett-Packard and Texas Instruments presently sell for about one-tenth of the original price of the HP–35 and TI Data Math. The present price of a four-function calculator is in the range of $4 to $7 and that of a simple (non-programmable) scientific calculator is in the range of $10 to $15. As a result, it can no longer be argued that calculators are too expensive for school students; even in school districts with large numbers of students from low-income households, it is possible for all students to have their own calculators as has been demonstrated by the Chicago Public Schools. This school district provides four-function calculators for students in Grades 4–6 and scientific calculators for students in Grades 7–9 (Dorothy Strong, personal communication).

It is still argued that calculators are not widely or effectively used in mathematics instruction. Overall this seems to be true (Kouba & Swafford, 1989, p. 102; Mathematical Sciences Education Board, 1989, p. 62), but there are beginning to be some good examples of ways in which mathematics can be taught and learned using calculators. Among these examples are:

1. The seventh- and eighth-grade supplementary materials *Getting Ready for Algebra* (Demana, Leitzel, & Osborne, 1988) developed by the Ohio State University Approaching Algebra Numerically project. Students studying from *Getting Ready for Algebra* use scientific calculators to learn about variables and their applications by employing strategies such as making tables and "guess-and-test."

2. The twelfth-grade textbook, *Transition to College Mathematics* (Demana & Leitzel, 1984) developed at Ohio State University for students who are classified as "remedial" by their scores on the placement test they were given as high school juniors by the Ohio Early Mathematics Placement Testing Program (Leitzel & Osborne, 1985). This text requires students to use scientific calculators.

3. The precalculus course developed by the Ohio State University Calculator and Computer Precalculus Project (Demana & Waits, 1989). Throughout this course students are expected to use a graphing tool; appropriate graphing calculators are made by Casio and by Sharp.

4. Two inservice teacher education modules developed by the Texas Education Agency; each of these modules is structured around the use of the TI Math Explorer, a fractions calculator, to teach fraction concepts and operations.

5. The materials being developed for Grades 7–12 by the University of Chicago School Mathematics Project (UCSMP). In each of the six courses under development calculators and computers are needed; in particular, in the fifth UCSMP course, Functions, Statistics and Trigonometry with Computers, a graphing unit (e.g., the Casio fx–7000G) is essential (Sharon Senk, personal communication).

Thus, it is presently possible—especially at the middle, junior high, and high school levels—to find teaching materials that require the use of calculators. These materials can either be used directly in classroom instruction or they can provide enough guidance so that teachers can create calculator-based student materials. Two significant problems seem to remain that prevent the widespread, effective use of calculators in mathematics instruction. One problem is that of training both preservice and inservice mathematics teachers so that they and their students can learn to use calculators effectively; this includes showing teachers effective ways of using calculators and persuading them that the use of calculators will improve, not diminish, students' abilities to learn mathematics and to solve problems. This is a significant problem in that large numbers of elementary and secondary school teachers teach mathematics and need to be trained; the magnitude of the problem is comparable to that faced when the New Math curricula were introduced in the 1950s and 1960s, since teachers need to learn both how to use a variety of calculators and how to explore the ways these calculators can be used to teach a mathematics curriculum restructured around the use of calculator and computer technologies.

The second, equally significant problem is the development of valid, reliable mathematics tests that require calculator use. At present, there are few established guidelines for the development of mathematics tests that require calculator use and only a scattered sample of nationally published mathematics tests for students who have and who know how to use calculators. The need to improve mathematics assessment in general and mathematics tests in particular to encourage the infusion of calculators in instruction (or in the classroom) is crucial. As *Everybody Counts* so succinctly states: "What is tested is what gets taught. Tests must measure what is most important" (Mathematical Sciences Education Board, 1989, p. 69). Thus, if we want to encourage teachers to use calculators in mathematics instruction, we must develop tests that require students to use calculators. This paper explores the guidelines by which such tests might be developed, and it cites a limited number of cases in which these kinds of tests have been developed and used.

CALCULATORS AND MATHEMATICS TESTING

In 1975, soon after the introduction of the TI Data Math and the HP–35, the National Advisory Committee on Mathematical Education (NACOME) urged that calculators be used in mathematics instruction (NACOME, 1975, pp. 40–43). In their conclusion on the advantages of using calculators in school mathematics instruction, they stated that "present standards of mathematical achievement will most certainly be invalidated by 'calculator classes'." A recommendation that calculators be used during mathematics instruction was made by the National Council of Teachers of Mathematics (NCTM) in its *An Agenda for Action*, which urges that "mathematics programs [should] take full advantage of calculators . . . at all grade levels" (NCTM, 1980, p. 1). In 1986, the NCTM again addressed the use of calculators in mathematics classrooms and specifically stated that

> The evaluation of student understanding of mathematical concepts and their application, including standardized tests, should be designed to allow the use of the calculator. . . . The National Council of Teachers of Mathematics recommends that publishers, authors, and test writers integrate the use of the calculator into their mathematics materials at all grades levels. (NCTM, April 1986)

In the interim, the College Board (1983; Kilpatrick, 1985), the Conference Board of the Mathematical Sciences (1983), the "School Mathematics: Options for the 1990s" conference (Romberg, 1984), and a joint symposium sponsored by the College Board and the Mathematical Association of America (MAA) (Kenelly, 1989) have all recommended that calculators be used during both mathematics instruction and mathematics testing.

Most recently, the NCTM Commission on Standards for School Mathematics (1989) based its recommendations on the assumption that all students will have a calculator available to them while studying mathematics. While this commission made no recommendation about the kinds of calculators that should be used in Grades K–4, they do recommend scientific calculators for middle school (i.e., Grade 5–8 students) and graphing calculators for students in Grades 9–12. Further, the Commission's first evaluation standard states that: "Methods and tasks for assessing students' learning should be aligned with the curriculum's instructional approaches and activities, *including the use of calculators*" [emphasis added] (NCTM Commission on Standards for School Mathematics, 1989, p. 193).

These recommendations have informed a broad audience, including mathematicians, mathematics educators and teachers, school administrators, and parents, that school and college mathematics curricula and tests of the future will require students to use calculators. However, neither singly nor in concert have the groups making the recommendations described the changes needed in present test and assessment procedures to assure that the achievement and aptitude of calculator-using students will be accurately measured. Because there have been no recommendations about the ways in which tests should be changed, three approaches have been used that permit students to use calculators while taking tests. These approaches

1. permit students to use calculators, but give them tests that make no provision for calculator use. I will call this approach *calculator-passive testing.*
2. permit students to use calculators, but give them tests developed so that none of their items require calculator use. This approach will be called *calculator-neutral testing.*
3. presuppose that students will need calculators while taking the test. The test is developed so that, for a majority of students, some portion of the items require calculator use in order to be solved successfully. An appropriate term for this approach is *calculator-based testing.*

In the next three sections, research and scholarship associated with each of these approaches will be considered.

Calculator-Passive Testing

This approach to the use of calculators during testing would require no changes in the mathematics tests that are presently administered to measure student achievement or aptitude. Letting students use calculators on tests that do not take into account or plan for that use is potentially hazardous since these tests may have on them items that are "calculator-unacceptable" (also called "calculator-sensitive). In a previous paper, I have defined *calculator-acceptable* and *calculator-unacceptable* items in this way:

1. An item is *acceptable* if (a) the objective(s) tested by it are the same whether or not a calculator is used and (b) the difficulty [level of the item] seems to be approximately the same when a calculator is used.

2.   An item is *marginally acceptable* if *only* 1a or 1b holds.

3.   An item is *unacceptable* if neither 1a nor 1b holds. (Harvey, 1989a, p. 28)

An item was judged to change in difficulty if the thinking level changes when a calculator is used while responding to the item; the thinking levels are those described by Epstein (1968):

1.   recall factual knowledge,
2.   perform mathematical manipulations,
3.   solve routine problems,
4.   demonstrate comprehension of mathematical ideas and concepts,
5.   solve nonroutine problems requiring insight or ingenuity, and
6.   apply "higher" mental processes to mathematics. (pp. 315–316)

Using these definitions, two placement tests that were part of the MAA Placement Test Program test package (i.e., Mathematics Test A/4A and Mathematics Test CR/1B) were studied. Mathematics Test A/4A examines knowledge of the content typically taught in basic, intermediate, and advanced algebra courses; Mathematics Test CR/1B tests skills and understandings prerequisite for calculus. On the thirty-two-item Mathematics Test A/4A, five of the items were judged to be marginally acceptable and six of the items to be unacceptable. On the twenty-five-item Mathematics Test CR/1B, there were three marginally acceptable and three unacceptable items. In Figure 8–1, three items from these two tests are shown; the first two were originally judged as marginally acceptable and the third as unacceptable (Harvey, 1989a, p. 29).

1. $\dfrac{2}{2 + \dfrac{1}{3}} =$

   (a) $\dfrac{3}{4}$ (b) $\dfrac{6}{7}$ (c) $\dfrac{4}{3}$ (d) 2   (e) 3

2. $(8^{-1/3})(9^{1/2}) =$

   (a) 6   (b) -6 (c) $(72)^{-1/2}$ (d) 2/3   (e) 3/2

3. For which values of x is tan x *not* defined?

   (a) $-\pi$ (b) $-\pi/2$ (c) 0   (d) $\pi/4$ (e) $\pi/3$

Figure 8–1.
Examples of marginally acceptable and unacceptable test items.

Using the above definitions of acceptability and unacceptability, I find all three items in Figure 8–1 unacceptable. When only paper and pencil are used, the first item tests students' knowledge of fractions concepts and operations and of the order in which operations are performed; when a calculator is used the only mathematical knowledge tested is order of operations. Item 1 is also less difficult if calculator use is permitted because students need only enter the numbers into the calculator in a correct order to obtain a decimal approximation of the correct answer; changing the distractors so that they are less familiar fractions closer to the correct answer might alleviate this problem.

The second item fails to test the objectives for which it was originally written (i.e., understanding of fractional and negative exponents) since, when a scientific calculator is used, the item tests the ability to enter the item stem into the calculator exactly as it appears except that parentheses are needed to enclose $-1/3$ and $1/2$. My present judgment is that this also reduces the difficulty level of this item.

Both the objective tested and the difficulty of the third item changes when a calculator is used because entering the numbers one by one and pressing the *tan* key will reveal the correct answer. Of the three items in Figure 8–1, this item would appear to be the most calculator sensitive.

The Effects of Calculator-Passive Testing. Several instances of calculator-passive testing have been reported. In six instances (Colefield, 1985; Connor, 1981; Elliott, 1980; Golden, 1982; Hopkins, 1978; Lewis & Hoover, 1981), standardized mathematics achievement tests were used. In three of these studies (Colefield, 1985; Hopkins, 1978; Lewis & Hoover, 1981), the scores of students who were permitted to use calculators were significantly higher than were the scores of those of who were not permitted to use calculators. A similar result was reported by Murphy (1981) who used the Problem Solving Achievement Test; the authorship of this test was not indicated in the dissertation abstract. Murphy reported that "students with unrestricted use of calculators achieved higher scores than students in the other three treatment groups in total problem-solving achievement." In this study there were two binary blocking variables (i.e., calculator use during instruction and calculator use during testing); crossing the values of these variables produced the four treatment groups.

Connor's study (1981) investigated the use of calculators during instruction on the concepts and techniques of trigonometry. One treatment group ($N = 48$) was permitted to use calculators during both instruction and testing; another treatment group ($N = 50$) was not permitted to use calculators during either instruction or testing. The test administered as the pre- and post-test in this study was the 1963 version of the Trigonometry Test in the Cooperative Mathematics Series. Analysis of the test data revealed no significant achievement differences between the two treatment groups.

Verbal problem solving was the focus of the study conducted by Elliott (1980). There were two treatment groups in this study: one group ($N = 70$) practiced verbal problem solving using only paper-and-pencil materials, while the second group ($N = 67$) was permitted to use calculators. Students in both treatment groups were given two post-tests; on one of these, they were permitted to use calculators and on the other, they were not. Elliott reported no significant differences between the treatment groups.

Golden (1982) studied the effects of calculator use on the achievement of EMR students in Grades 7–9. There were two treatment groups; one group ($N = 23$) used calculators while studying the four fundamental algorithms while the other group ($N = 27$) did not. On the post-test there were no significant differences between the two groups on addition and subtraction items. The calculator-using students did perform significantly better ($p < 0.05$) than the other group on multiplication and division items.

There is only one calculator-passive study that attempted to discover the effects of using calculators on an existing test. Gimmestad (1982) randomly chose a group of nineteen students from among those taking a Calculus II course at Michigan Technological University; all of the students in this group had been permitted to use calculators in the course. Nine of the students composed the calculator group; the remaining ten students, the non-calculator group. Each student was asked to "think aloud" while solving twenty-four sample problems from the College Board's Advanced Placement Calculus Examination. Each student interview was videotaped, coded, and ana-

lyzed for the reasoning processes used and the results produced by the student. These outcomes are important:

1. Calculator use seemed to change the strategies
    used to solve only "a couple of problems." Gim-
    mestad does not identify which problems were
    solved with changed strategies, but she does com-
    ment that multiple-choice problems in which the
    checking method strategy (Harvey, 1989a) can be
    employed seemed to be calculator-sensitive. The
    use of this process was negatively correlated with
    the product scores ($r = -0.25$).[1]

2. Exploratory manipulations were more effective
    when calculators were used. The problem cited by
    Gimmestad was one that involved finding a limit.
    The process variable *manipulation* was also nega-
    tively correlated with product score ($r = -0.13$).

3. The frequency of checking by retracing steps for
    the calculator-using students was twice that of
    students not using calculators. The correlation re-
    ported for this process variable with the product
    score is 0.47. Based on this result, Gimmestad
    concluded "this may be an important difference
    between testing calculus with and without the cal-
    culator" (p. 3).

4. There was no significant difference between the
    mean product scores of the test group that used
    calculators and the group that did not.

With the exception of Gimmestad's study, none of the calculator-passive studies reported here attempted to discover how
the use of calculators during testing changed the processes
used by students or the objectives that were tested. In the
other studies, there seems to have been an implicit assumption
that the objectives tested by an item remained unchanged when
calculator use was permitted. This assumption permitted Lewis
and Hoover (1981) to argue, based on their results, that since

---

[1] The correlations reported are between the use of processes and the
product score for all nineteen of the students and not for the nine
calculator-using students. I speculate that these correlations might
have been different if the process use and product scores of only the
calculator-using students had been used.

there was a nearly perfect correlation between the ranks of the students on the regular and calculator administrations of the test they used, the pupil percentile ranks would be the same whether or not calculators were used. Thus, they concluded that the only change that would be needed to permit the use of calculators on this standardized test would be to re-norm the test using data from calculator administrations of it.

However, as I have already argued, item objectives can change when calculators are used, especially on computational items, since these items can be answered by simply keying into the calculator an appropriate sequence of numbers and operations. As a result, at least the "strictly" computational items on standardized tests are no longer testing mathematics achievement but instead are testing students' calculator facility; the result is a changed—and possibly distorted—picture of a student's mathematics achievement or aptitude.

Calculator-Neutral Testing

Calculator-neutral tests are tests that permit but do not require those taking them to use calculators. To achieve this goal, the tests must not include any items on which calculator use will benefit test takers. One way in which calculator-neutral tests have been developed is to begin with an existing test and to determine, in some way, which of its items are calculator-sensitive. Once this determination has been made, the calculator-sensitive items are replaced with new items that are not calculator-sensitive (Leitzel & Waits, 1989). A similar strategy may have been used by persons developing new calculator-neutral tests, but in the studies I have examined (Abo-Elkhair, 1980; Casterlow, 1980; Long, Reys, & Osterlind, 1989; Mellon, 1985; Rule, 1980) the ways in which the items were generated in order to make them calculator neutral were not discussed.

The National Assessment of Educational Progress (NAEP) (1988) has defined a calculator-neutral item as one whose solution does not require the use of a calculator (p. 33). Ideally, a calculator-neutral item should be one to which calculator-using and non-calculator-using students respond equally well. So, when applying the NAEP definition to a potential test item both the objectives being tested and any calculator-related skills needed have to be considered. Figure 2 shows what I consider to be a stereotypical calculator-neutral item adapted from an item on the calculator-neutral test given by Abo-Elkhair (1980).

> The number of students in five classes is 25, 21, 27, 29, and 28. What is the average number of students in each class?

Figure 8–2.
Stereotypical calculator-neutral test item.

The study in which this item was used (Abo-Elkhair, 1980) was one in which students were taught about averages and averaging. This item is calculator-neutral for students with a good understanding of the paper-and-pencil algorithms for addition and division since they will not need a calculator to perform the necessary computations; the responses of these students should accurately reflect their understandings of averaging. However, when students not having the needed computational proficiency are tested, the item may not help to accurately measure student understanding. Within this group, the responses of the students who have facility with and are permitted to use calculators will depend upon their understandings of averages and averaging. But if students in this group do not know how to use or are not permitted to use calculators, then their responses will not reflect their knowledge of the objective being tested. This effect could be eliminated by providing all students with calculators and helping them to acquire the needed calculator-related skills. However, this has not typically been the practice; in all of the studies I examined, both calculator-using and non-calculator-using students participated. Perhaps the most extensive examination of calculator-neutral testing has been reported by Leitzel and Waits (1989).

The test used by the Ohio Early Mathematics Placement Testing Program for High School Juniors (EMPT) is calculator-neutral (Leitzel & Waits, 1989). Until 1983–84, students taking the thirty-two-item EMPT test were not allowed to use calculators; since then students have been permitted to use any calculator they bring with them to the testing. Essentially, the same thirty-two items were used on EMPT tests EB1 and EB2 during the school years 1979–80 through 1982–83. In 1983–

84, a test consisting of six new items and twenty-six items from the EB2 test was given. During the next two years, three items were replaced to produce the test, EB4/5, used since 1984–85. The EMPT test EB4/5 has twenty-three items in common with EB1 and EB2. In modifying EB1 and EB2 to obtain EB3 and its successor EB4/5, only one item was eliminated because it was "obviously calculator-dependent." (p. 18).

Leitzel and Waits reported that the means on EMPT tests were fairly stable from 1977–78 to 1985–86; during this period the means varied from a low of 14.2 to a high of 15.2. Data reported for the school years 1983–84, 1984–85, and 1985-86, when students were permitted to use calculators, showed that during each of these years, calculator-using students had higher test means than did non-calculator-using students. Unfortunately, Leitzel and Waits neither reported nor statistically compared the means of the two groups of students. They have, however, studied some of the characteristics of calculator-using versus non-calculator-using students. The calculator-using students were more likely than their non-calculator-using peers to: (a) be planning to attend a four-year college, (b) be taking Algebra II as a junior, (c) be taking advanced mathematics courses as a junior, and (d) have made a grade of *A* or *B* in the last mathematics course they took. Based on these data, Leitzel and Waits concluded that "it is not surprising that the group using calculators performed at a higher level."

These investigators have also examined the difficulty levels and calculator sensitivity of their test items. For most of the ten easiest and ten most difficult items on the EB4/5 test, both calculator-using and non-calculator-using students found the items to be almost equally difficult. Only one of the items seemed to be much less difficult for the calculator-using students; that item, Item 8, is shown in Figure 8–3. It also proved to be the most calculator-sensitive.

The decimal fraction 0.222 most nearly equals:

(A) 2/10   (B) 2/11   (C) 2/9   (D) 2/7   (E) 2/8

Figure 8–3.
EMPT test item that was less difficult for calculator-using students.

To examine the calculator sensitivity of their items, Leitzel and Waits (p. 22) developed a calculator sensitivity (CS) index that is defined as follows:

$$CS = \frac{\% \text{ correct by calculator-using group}}{\% \text{ correct by non-calculator-using group}}$$

The thirty-two items on the EMPT test EB4/5 have CS values ranging between 1.1 and 2. The higher the CS index, the more calculator-sensitive the item would seem to be. Three EMPT items had a CS index greater than 1.6; six had CS indices between 1.5 and 1.6; and the remaining twenty-three items had CS indices not greater than 1.5. Leitzel and Waits "believe the expected CS index range for items [on EB4/5] that are not calculator-sensitive is between 1.5 and 1.2."[2]

The three items with the lowest CS indices require that students (a) find a simultaneous solution for a pair of linear equations, (b) simplify the sum of two rational functions, and (c) multiply a quadratic expression by a linear one. The three items with high CS indices are Item 8 (see Figure 8-3) and items that ask students to (a) compute the value $(1/2)^{-3}$, and (b) simplify $\sqrt{32} - \sqrt{2}$. When judged by the criteria for calculator-acceptable and calculator-unacceptable items given earlier in this chapter, I judged the three items with low CS indices to be acceptable and the three items with high CS indices to be unacceptable.

A less extensive study than that by Leitzel and Waits (1989) was the study by Long, Reys, and Osterlind (1989) who investigated the differences in the scores of calculator-using and non-calculator-using students in Grades 8 and 10 on the Missouri Mastery and Achievement Tests (MMAT). The investigators reported that the tests for Grades 7–10 were not designed to test calculator use (i.e., were intended to be calculator-neutral). Even so, of the nine released items from the eighth-grade and tenth-grade MMAT tests, seven would seem to be calculator-sensitive even for students using a four-function calculator. Forty-five percent of the eighth-graders and 56 percent of the

---

[2] I hypothesize that the lower bound on CS indices should be 1.00 instead of 1.2. If CS is an accurate measure of calculator-sensitivity, then the best calculator-neutral item would be one with a CS index of 1.00.

tenth-graders who took MMAT tests reported that they used a calculator. At the eighth-grade level, calculator-using students significantly outperformed non-calculator-using students on the test and on three of the four MMAT subtests: (a) understanding numbers, (b) computation, and (c) interpretation and application ($p < 0.001$). At the tenth-grade level, the calculator-using students also significantly outperformed non-calculator-using students on the total test and on two of its three subtests: (a) computation and (b) interpretation and application ($p < 0.001$).

The outcomes described by Long, Reys, and Osterlind (1989) are like those reported by Abo-Elkhair (1980), Casterlow (1980), and Mellon (1985). Rule (1980) does not report significant differences between his two treatment groups (calculator versus non-calculator), who studied functions, graphing, function composition, and inverse functions. On Rule's calculator-neutral tests, calculator use would be of little or no assistance; the items require students to manipulate symbolic expressions, interpret graphs, or generate graphs. In one sense, on a calculator-neutral test like Rule's, I would not expect to find differences between calculator-using and non-calculator-using students since the calculator-using students really have no opportunity to use calculators while taking the test. On the other hand, it is disappointing that Rule did not find differences between his two treatment groups, since the commonly expressed hope is that calculator-using students will develop better conceptual understanding of the mathematics they study. However, the instructional part of Rule's study lasted for only eleven consecutive instructional days, and the lessons presented do not include explorations of the kind that may be needed to produce deeper conceptual understanding. So, it is possible that Rule's calculator-neutral test would have shown the same kinds of differences described by the other studies discussed had the calculator-using students had more opportunities to explore functions using their calculators as tools.

Overall, the uses of the calculator-neutral tests described here showed that calculator-using students more often than not outperformed their non-calculator-using counterparts. These studies also show that great care must be used in developing calculator-neutral items that *might* permit calculator use; in some instances, lack of rigor in developing these items can result in an inaccurate test of the objectives stated for the item or in an item that is calculator-sensitive instead of being calcu-lator-neutral. It seems to me that valid calculator-neutral tests can be developed that can be used with both calculator-using and non-calculator-using students; the data presented here show, however, that it may be necessary to norm the scores from these two groups separately.

Calculator-Based Testing

Until all mathematics students have and know how to employ calculators, it will probably be necessary to develop and use calculator-neutral tests. However, when all students have avail-able and can use calculators effectively as tools, it will be pos-sible to administer calculator-based tests routinely.

In the course of my work, I define calculator-based math-ematics tests and calculator-active test items as follows:

> A *calculator-based mathematics test* is one that (a) tests mathematics achievement, (b) has some calculator-ac-tive test items on it and (c) has no items on it that could be, but are not, calculator-active *except for* items that are better solved using non-calculator based techniques. A *calculator-active test item* is an item that (a) contains data that can be usefully explored and manipulated us-ing a calculator and (b) has been designed to require active calculator use. (Harvey, 1989b, p. 78)

These definitions must be interpreted by those using them. The first two criteria for a calculator-based test can be strictly ap-plied. The first is intended to affirm that the objectives of calcu-lator-based mathematics tests should be mathematics objec-tives and that it is not the intent of these tests or their items to test calculator facility solely. Criterion (a) proceeds from the assumption that test takers will already have adequate facility with the calculator to take the test; this criterion agrees with a recommendation made by a joint symposium on the use of calculators in standardized testing convened by the College Board and the Mathematical Association of America (Kenelly, 1989, p. 47). Criterion (b) is intended to ensure that a calcula-tor-based test will contain items that require students to use their calculators while taking the test; this criterion distin-guishes calculator-based tests from calculator-neutral tests like the one given by Rule (1980). The third, criterion (c), cannot be strictly applied since a judgment is required about the best way of solving a problem. As an example, consider the items shown in Figure 8–4. These items appear on college-level placement tests being developed by the MAA Calculator-Based Placement Test Program Project.

$$\frac{x^2 - 9}{2x} \cdot \frac{6}{3x + 9} =$$

    a) −1   b) −3   c) $x - 3$ d) $\dfrac{x + 3}{x}$ e) $\dfrac{x - 3}{x}$

If $2^{3000} = 10^x$, then x is

    a) 600   b) 903.090   c) 2079.442   d) 9965.784   e) undefined

Figure 8–4.

Test items satisfying criterion (c) of a calculator-based test.

There are at least two ways to solve the first problem; one of these ways would seem to require calculator use while the other does not. On an untimed test, calculator use would permit use of the specific-instance strategy (Harvey, 1989a); to employ this strategy a student would substitute numbers for the variable x in each of the expressions in the item stem and in the foils and would select an answer by comparing the numeric evaluations of the item stem to those of the foils. A second, and better, way of solving this problem is simply to factor each of the expressions in the item stem, cancel like terms wherever possible, and combine the remaining terms so as to reach one of the multiple-choice answers.

A mathematically correct way to solve the second problem would be to find 2^{3000} and then to take the base-10 logarithm of that result. Present calculators give an error message when 2^{3000} is entered. A way to solve this problem is first to take the base-10 logarithm of both sides of the equation and then to multiply log (2) by 3000. Another way to solve the problem would be to estimate the size of 2^{3000} as 10^{1000} and so, to determine that x is about 1000. Using this approximation, the correct answer to the problem becomes apparent.

Students taking calculator-based tests will be "calculator-dependent"; that is, they will actively use calculators as tools and many of the techniques and algorithms they use for solving problems will be calculator-based. Students who study from curricula that meet the NCTM *Standards* (NCTM Commission on Standards for School Mathematics, 1989) will be calculator-dependent. Criterion (c) was included in the definition of calculator-based mathematics tests to require test developers to investigate the calculator solutions for each item and to judge whether or not a calculator-based solution can and should be expected. This examination should produce valid tests for calculator-dependent test takers. This criterion is also intended, for the present, as a reminder to experienced mathematics test developers that the non-calculator techniques and algorithms for solving problems that they know and successfully apply are not the ones that will be used by the intended test audience.

My definition of a calculator-active test item is like that given by the National Assessment of Educational Progress (1988, p. 33). However, the definition given here is more specific in that it insists that the item contain data that can be usefully explored using a calculator. This criterion is intended to prevent labeling an item as calculator-active when, for example, the only calculator activity required is to change an answer computed as a common fraction into a decimal approximation of that answer. The second criterion for a calculator-active item should probably be modified to read, "designed so as *most likely* to require calculator use," since there are presently few instances in mathematics where paper-and-pencil procedures cannot be used. For example, it is not easy without a calculator to approximate the powers of e or continuously compounded interest or to compute the combinations of n objects taken r at a time when n is large, but there are paper-and-pencil techniques that apply in each situation. However, it is not likely that calculator-dependent test takers would think to use these techniques—even if they know about them—because the calculator is more facile and faster in these situations. Even with the suggested modification, the second criterion signals test developers that they must plan for active calculator use and, if the item is a multiple-choice one, to develop the foils based both on the mathematical errors students make while solving such problems and on the exact form those incorrect answers take when calculators are used.

Recommendations for Test Construction. The joint symposium sponsored by the College Board and the Mathematical Association of America considered and made recommendations about a number of issues related to the development of calculator-based mathematics tests. One of these recommendations, that the tests be curriculum based and should not measure *only* calculator skills or techniques, has already been mentioned. The symposium participants made eight additional recommendations; three of these recommendations pertain to the development of calculator-based mathematics tests. The pertinent recommendations are:

1. Studies are needed that will identify the content areas of mathematics that have gained in importance because of the emergence and use of technology. In addition, the ways in which achievement and ability are measured in these areas should be studied as should new ways of *testing* achievement and ability.

2. Choosing whether or not to use a calculator when addressing a particular test question is an important skill. Thus, not all questions on calculator-based mathematics achievement tests should require the use of a calculator.

3. Nationally developed tests of calculator-based mathematics achievement tests should provide descriptive materials and sample questions that clearly indicate the level of calculator skills needed. (Kenelly, 1989, pp. 47–48)

The last cited recommendation goes on to say that students should be permitted the use of any calculator as long as it has the functionality required to solve the problems on the test. When these two parts are taken together the result is a recommendation that test developers specify the least capable calculator needed to respond to the calculator-active items successfully but should not bar the use of more capable calculators. At the time of the symposium, the first graphing calculator, the Casio fx-7000G, had just been introduced and the Hewlett-Packard HP28C would not be introduced for another three months. These calculators and calculators like them can give students who have them an advantage. Graphing calculators add the possibility of geometric problem solving; calculators that graph and symbolically manipulate might change some test items into tests of the student's calculator facility.

For example, if asked to find the real zeros of a polynomial function, a student having only a scientific calculator could use that calculator to check for the rational zeros of the function, to develop a table of values so as to sketch a graph of the function, and to apply numeric techniques for approximating the real zeros. A student solving that problem with a graphing calculator could develop a complete graph of the function, observe the places where that graph crosses the x-axis, and use the calculator's "zoom-in" program or [SOLVE] key to approximate the real solutions. Even if these two students were asked to find only one real zero of a given polynomial function, the student having the graphing calculator would still have an advantage. Thus, in developing calculator-based tests it will be necessary to specify both the least capable and the most capable calculator that can be used while taking the test.

Research on Calculator-Based Mathematics Tests. It seems quite likely that many calculator-based tests have been developed; however, not many of these tests have been widely circulated or discussed. As a result, this section discusses a single dissertation study, the tests being developed by the MAA Calculator-Based Placement Test Program Project, and the chapter tests developed by the Ohio State Calculator and Computer Precalculus Curriculum (C²PC) Project.

The three calculator-based unit tests were developed by Bone (1983) as part of her study of the effectiveness of an introductory unit on circular functions; all of the items on these tests were free response items. The total number of questions and calculator-active questions on each test is given in Table 8–1; Bone did not report test and item statistics for any of the tests. The four calculator-active items on which Bone's ten subjects scored most poorly were two word problems, an item that asked students to make a table of values in order to sketch the graph of a function, and an item that asked students to find the value of an angle in the third quadrant given its cosine. The calculator-active items that almost all students

Table 8–1
Types of Questions on the Tests Developed by Bone

Unit Test	Number of Questions	Number of Calculator-Active Questions
I	24	14
II	46	10
III	26	1

Note. Source: Bone, 1983. Used with permission.

answered correctly are on the Unit I test and are items requiring simple computations. Examples of the test questions used by Bone are in Figure 8–5.

Unit I Test Items

1. Use a calculator to find the value of *sin* 27° to 4 decimal places. (2.5 points)

2. Change 4.75 from radians to degrees (to the nearest 0.001). (2.5 points)

3. Assuming the earth is a sphere with radius of 4000 miles, how far is Tokyo from Adelaide (to the nearest mile)? Tokyo, Japan, is located 35° 30' North latitude and Adelaide, Australia, is located at about 35° South latitude, just about due south of Tokyo. (10 points)

Unit II Test Items

4. Use a calculator to find the value of *cos* (–11/6) to 4 decimal places (to the nearest 0.001). (2 points)

5. Given *sec* θ = 2.13 find the smallest positive measure for θ to the nearest degree. (2 points)

Unit III Test Item

6. Sketch the graph of *y* = 3 *sin* (4π*t*) for 0 ≤ *t* ≤ 1. Locate points at intervals of 0.25. (6 points)

Figure 8–5.
Examples of calculator-active items included on Bone's unit tests.
(Bone, 1983). Used by permission.

The MAA Calculator-Based Placement Test Program (CBPTP) Project is developing six college-level placement calculator-based placement tests and two high school prognostic calculator-based tests intended to be used in testing high school juniors in order to forecast the mathematics courses these students would be placed in upon entering college if they take no additional high school mathematics courses. The CBPTP college-level placement tests will all require students to use scientific (non-programmable) calculators; the high prognostic tests will require students to use graphing calculators. On each of these six tests, about 25 percent of the test items will be calculator-active.

Development of two of these tests, the Calculator-Based Arithmetic and Skills Test (CB-A-S) (Boyd et al., 1989) and the Calculator-Based Calculus Readiness Test (CB-CR) (Kenelly et al., 1990), has been completed. Development of two additional tests, the Calculator-Based Basic Algebra Test (CB-BA) and the Calculator-Based Algebra Test (CB-A), is nearly completed. The data reported here are from the tryouts of these tests with high school or college students.

The CB-A-S test consists of thirty-two items that test students' knowledge of arithmetic and pre-algebra; seven of the items on this test are calculator-active. When administered to 191 students enrolled in remedial college mathematics courses, the mean score on this test was 16.01 (s.d. 6.53), the reliability (coefficient α) was 0.86, and the mean item difficulty was 0.50. The r-biserials and item difficulties for the seven calculator-active items are shown in Table 8–2.

Table 8–2
r-biserials and Difficulties for CB-A-S
Calculator-Active Items

Item	r-biserial	Item difficulty
1	0.44	0.30
2	0.44	0.18
3	0.54	0.60
4	0.45	0.55
5	0.57	0.41
6	0.46	0.34
7	0.56	0.30

Note. Source: Mathematical Association of America, 1989, 1990. Used by permission.

All of the calculator-active items on the CB-A-S test correlate well with the other items on the test; most of the items are, as intended, of medium difficulty though, in general, they are harder than is the average item on this test. The hardest item,

Item 2, is one that asked for an approximation of the value of $N^2 - N$ when $N = -1.1$. The next two most difficult items asked students to approximate the value of $(1 + 1/6)^4$ and of $1/(3 + \sqrt{5})$, respectively. The easiest calculator-active item, Item 3, asked for an approximation of n when $n \times n \times n = 63$.

The CB-CR test has two parts. The first part is composed of twenty items intended to test precalculus knowledge; the second, five-item part specifically tests knowledge of trigonometry and elementary functions. There are nine calculator-active items on CB-CR. When given to a group of thirty-six high school seniors finishing their precalculus course, the mean score on this test was 16.81 (s.d. 3.49), the reliability (coefficient α) was 0.65, and the mean item difficulty was 0.67. When the calculator-active items are considered alone the mean score was 4.67 (s.d. 2.11), the reliability (coefficient α) was 0.61, and the mean item difficulty was 0.52. The r-biserials and item difficulties for the nine calculator-active items are shown in Table 8–3.

Table 8–3
r-biserials and Difficulties for CB-CR
Calculator-Active Items

Item	r-biserial	Item difficulty
1	0.50	0.56
2	0.44	0.50
3	0.18	0.97
4	0.44	0.53
5	0.46	0.36
6	0.55	0.39
7	0.45	0.39
8	0.28	0.50
9	0.48	0.47

Note. Source: Mathematical Association of America, 1989, 1990. Used by permission.

The item that is easiest is also the item with the lowest r-biserial. This item defines two functions and asks that a value of the function that is the composition of the given functions be computed at $x = 1.7$. The other item whose r-biserial value is an outlier is Item 8; this item requests that the value of $\tan(2\pi/5)$ be computed. The remaining calculator-active items are of medium difficulty and correlate well with the other, non-calculator-active items on the test that were, for the most part, taken from the existing MAA calculus readiness placement test.

The Calculator-Based Basic Algebra Test (CB-BA) (Curtis et al., in press) consists of twenty-five items from basic and intermediate algebra. On the CB-BA test, eight of the items are calculator-active. When this test was given to 256 students completing their algebra and precalculus courses, the mean score was 16.44 (s.d. 3.90), the reliability (coefficient α) was 0.75, and the mean item difficulty was 0.66. On the subtest consisting of the calculator-active items, the mean score was 4.08 (s.d. 1.64), the reliability (coefficient α) was 0.38, and the mean item difficulty was 0.51. For those students who are planning to take courses beyond basic and intermediate algebra, the test proved to be slightly more difficult than is usually intended for placement tests; however, the more difficult items would appear to be among the calculator-inactive and calculator-neutral items. These items were largely drawn from among those on the existing MAA Basic Algebra Test. The r-biserials and item difficulties for the calculator-active items is shown in Table 8–4.

Table 8–4
r-biserials and Difficulties for CB-BA
Calculator-Active Items

Item	r-biserial	Item difficulty
1	0.35	0.63
2	0.38	0.33
3	0.42	0.71
4	0.23	0.39
5	0.24	0.34
6	0.37	0.45
7	0.48	0.43
8	0.29	0.79

Note. Source: Mathematical Association of America, 1989, 1990. Used by permission.

Items 4 and 5, the two items on CB-BA with the lowest r-biserials, also are two of the harder calculator-active items on this test. In each case, the two lowest quintiles of students taking the test responded correctly to these items less than 30 percent of the time, and the highest quintile responded correctly to the item about 60 percent of the time. One of these items asked students to determine the interval in which the graph of $2.5x - y + 8.2 = 0$ crosses the x-axis, while the other asked students to approximate the larger root of a quadratic equation. The two easiest calculator-active items, Items 3 and 8, asked respectively for an approximation of x when $(x - 5)^5 = 10$ and for the missing test score given three of the scores and the mean score.

The remaining test that has been developed, so far, by the Calculator-Based Placement Test Project is the Calculator-Based Algebra Test (CB-A) (Cederberg et al., in press). The CB-A test includes items from basic, intermediate, and college algebra. This test consists of thirty-two items; eight items are calculator-active. At present, data are available for only six of the eight calculator-active items, because after the last tryout of the test ($N = 210$), two of the items were discarded and replaced with new ones. When these two items were deleted and the data from the last tryout were reanalyzed on the resulting thirty-item subtest, the mean score was 11.56 (s.d. 3.49), the reliability (coefficient α) was 0.51, and the mean item difficulty was 0.39. The mean score on the six calculator-active items was 1.73 (s.d. 1.15); this subtest had a reliability (coefficient α) of 0.11 and a mean item difficulty of 0.29. In contrast, the twenty-four-item subtest consisting of the twenty-four calculator-inactive and calculator-neutral items produced a mean score of 9.82 (s.d. 3.10), a reliability (coefficient α) of 0.50, and a mean item difficulty of 0.41. Overall, this was a difficult test for the sample of students who took it, and for those students the calculator-active items were, overall, more difficult than were the non-calculator-active items. The r-biserials and item difficulties for the calculator active items are shown in Table 8–5.

Table 8–5
r-biserials and Difficulties for CB-A
Calculator-Active Items

Item	r-biserial	Item difficulty
1	0.29	0.41
2	0.12	0.30
3	0.36	0.27
4	0.12	0.23
5	0.13	0.19
6	0.19	0.32

Note. Source: Mathematical Association of America, 1989, 1990. Used by permission.

Items 2, 4, and 5 have low r-biserials; two of these items are the most difficult of the calculator-active items on the test. Item 2 is a similar triangle problem in which all of the lengths given are decimal fractions. Item 4 seeks an approximation to the root of a quadratic equation, and Item 5 seeks a similar approximation, except that the equation in the item stem is $\sqrt[3]{x^2 + 1} = 2.98$. It is my observation that these are concepts and skills that are usually difficult for students. Thus, the use of a calculator to solve these problems may contribute to their difficulty but calculator use is not a major factor.

Since the content of the four calculator-based placement tests range from tests of arithmetic and skills to knowledge of precalculus, the calculator-active items suggest that calculator use should be expected during their solution. Figure 8–6 shows representative calculator-active items that appear on CB-A-S, CB-BA, CB-A, or CB-CR. The distractors for each of these items are based upon mathematical *errors* that students make and are not intended to measure students' calculator facility. It is expected that students who take these placement tests will already have and know how to use calculators of the kind needed.

1. Which of the following best approximate $6(1.4 - 1.2)^5$?

   (A) 0.0003   (B) 0.0019   (C) 0.7805   (D) 0.7850   (E) 17.3395

2. The approximation of $(1 + 1/6)^4$ correct to 4 decimal places is

   (A) 1.0008   (B) 1.1667   (C) 1.8526   (D) 2.1614   (E) 4.6667

3. If $x^3 + 2.75 = 5.12$, then which of the following best approximates x?

   (A) 1.33   (B) 13.31   (C) 113.42   (D) 131.47   (E) 487.44

4. The radius of the larger of two concentric circles is 6.9; the radius of the smaller circle is 4.7. Which of the following numbers best approximates the area of the region between the two circles?

   (A) 13.82   (B) 15.21   (C) 25.52   (D) 80.17   (E) 149.57

5. A pole 7.8 feet high casts a shadow 12.8 feet long. If the length of a shadow cast by a tree is 83.9 feet, which of the following best approximates the height of the tree?

   (A) 51.1   (B) 78.9   (C) 88.9   (D) 99.8   (E) 137.7

6. Which of the following best approximates a solution of $x^2 - 4x = 3$?

   (A) −2.65   (B) −0.65   (C) 0.65   (D) 1.73   (E) 3

7. Which of the following best approximates the number approached by the sequence $(3/2)^4$, $(4/3)^6$, $(5/4)^8$, ... $((n + 1)/n)^{2n}$, ...?

   (A) 1   (B) 2.718   (C) 6.192   (D) 7.389   (E) No finite number

8. Which of the following best approximates the sum of the areas of the rectangles shaded in the figure below?

(A) 0.131
(B) 0.163
(C) 0.194
(D) 0.538
(E) 1.944

Figure 8–6.
Sample calculator-active items.

While the four tests developed by the MAA Calculator-Based Placement Test Program Project give examples of calculator-active items that can be developed when scientific calculators are required, these tests provide no examples of the kinds of items that result when graphing calculators are required. The Ohio State University Calculator and Computer Precalculus Curriculum (C^2PC) Project has developed precalculus text materials that require the use of calculator or computer graphing tools (Demana & Waits, 1989) and tests; some of whose items also require the use of these same tools (Demana et al., 1990). Items from these tests were used on the two midterm tests and the final examination in a single section of Algebra and Trigonometry at the University of Wisconsin-Madison during the Fall Semester, 1989–90. Table 8–6 describes some characteristics of these tests.

Table 8–6
Characteristics of Algebra and Trigonometry Tests

Test	Number of students	Free-response items	Multiple-choice items			
			N	Mean (s.d.)	Reliability	Difficulty
Mid-term I	75	6	26	64.16[a] (7.94)	0.52	0.82
Mid-term II	72	6	26	58.67[a] (11.02)	0.73	0.75
Final exam	72	11	19	36.00[a] (9.05)	0.63	0.63

[a]Each item had a value of 3 points.

On the first midterm test, five free-response and three multiple-choice items clearly would require students to use the graphics capabilities of their calculators to respond to them. On the second midterm test, these numbers were seven and four, respectively. On the final examination, only one multiple-choice item required graphics calculator capability; a majority of the free-response portion of this test required students to use graphics capabilities. Table 8–7 gives the r-biserials and item difficulties for the eight multiple-choice items on the tests that were graphics calculator-active.

Table 8–7
r-biserials and Difficulties
Graphics Calculator-Active Item

Item	r-biserial	Item difficulty
1	0.26	0.87
2	0.46	0.87
3	0.35	0.85
4	0.48	0.57
5	0.43	0.60
6	0.30	0.25
7	0.41	0.88
8	0.36	0.67

Most of the graphing calculator-active multiple-choice items contain specific instructions to use a graph, or implicitly suggested that use. An example of an item that specifically told students to use a graph is the one on the first mid-term test that stated: "Use a graphing utility to determine the number of real solutions to the equation $4x^3 - 10x + 17 = 0$." On that same test the item stem implicitly called for the use of a graphing utility when it asked: "Which one of the following viewing rectangles[3] gives the best complete graph of $y = 10x^3 - 6x^2 + 20$?" Overall, these items correlated satisfactorily with all of the multiple-choice items on the test; they ranged from very easy to moderately difficult.

On each of the three tests administered to the algebra and trigonometry class there were a number of items that could be solved algebraically or graphically. It is not known how stu-

---

[3] The *viewing rectangle* is a description of the minimum and maximum x- and y-coordinates that are shown on the graphics screen.

dents *actually* solved these problems; an example was the one that asked students to determine the period of the function $f(x) = 3 \sin(\pi x)$.

The free-response items on these tests varied in that some of them requested symbolic manipulations and solutions, some expected students to use algebraic techniques and algorithms to produce exact solutions, and some required the use of graphing calculators for their solution. The scores on the free-response and the multiple-choice portions of the two midterm tests were moderately well correlated. The correlation coefficient of these two parts on the first midterm test was $r = 0.63$; it was $r = 0.61$ on the second midterm test. On the first test, students correctly responded, on average, to 82.18 percent of the multiple-choice items and to 65.48 percent of the free-response items. On the second midterm test, the corresponding data were 75.21 percent and 69.36 percent, respectively. The corresponding data for the final examination have not yet been computed.

The tests and test items that have been produced by the MAA Calculator-Based Placement Testing Project and the Ohio State C²PC Project demonstrate that valid, reliable calculator-based tests and calculator-active items can be generated that satisfy the definitions of these terms that were given earlier in this paper. At present there is a paucity of published calculator-based tests and calculator-active items. In order to study the items that have been developed and, at the same time, to keep them secure, faculty from the University of Chicago, the Ohio State University, the Pennsylvania State University, and the University of Wisconsin-Madison are establishing a pool of calculator-active items in content areas including algebra, precalculus, and functions.

CONCLUSION

This paper begins with a quote that avers that present testing practices hold today's students hostages to yesterday's mistakes. One reason for this is that "What is tested is what gets taught" (Mathematical Sciences Education Board, 1989, p. 69). Thus, as long as mathematics tests fail to incorporate the use of calculators, I am certain that mathematics instruction will fail to incorporate the use of calculators effectively, and so today's students will be prisoners to a mathematics curriculum that is failing to prepare them for the society in which they will live both now and in the twenty-first century.

Just permitting students to use calculators while taking mathematics tests will not be enough. Students will need to be taught how and when to use calculators while solving all kinds of mathematics problems. Equally important, tests will have to actively account for the changes in the ways that mathematics problems are solved and the kinds of mathematics problems that can be solved when calculators are used. I conclude that calculator-passive and calculator-neutral tests do not satisfactorily account for these changes and that only calculator-based tests can. In addition, it seems clear that each time calculators become more capable and more responsive to mathematics instruction—and each is occurring—mathematics tests will have to be changed.

While the use of calculators on mathematics tests and, by implication, in mathematics instruction will not remedy all of the failings of present tests and instruction, their use is necessary if we want students to investigate, to explore, and to discover mathematics.

9

Gender Differences in Test Taking:
A Review

Margaret R. Meyer

Ideally, when students take a mathematics exam, the only thing that should influence their score is their mastery of the material being tested. This paper reviews evidence concerning gender differences in mathematics test taking. It examines several factors which have surfaced relating to differences in performances for males and females. One conclusion reached is that the use of the multiple-choice format may result in a male advantage. A recommendation is therefore made that assessment instruments be developed that do not rely as heavily on the multiple-choice format.

Do males differ in their mathematics test-taking performance independent of their understanding of the mathematics being tested? This review attempts to answer this question. Although the focus will be on mathematics tests, very little research has looked specifically at mathematics test taking. Therefore, evidence from more general test taking will be presented.

Several factors have been investigated that relate to differences in test performance for males and females. These factors are power vs. speed test conditions, item difficulty sequencing, exam format, test-wiseness, risk taking behavior, and test preparation behaviors. The first three of these factors have received

169

the most attention in the literature. The other factors are usually included in studies as covariates. Gender of examinee is
not always included as a factor of interest. Those studies that
did not include gender will be reviewed only when they illumi-
nate those that did include it.

POWER VS. SPEED TEST CONDITIONS

One characteristic of a test is the amount of time available to
complete it. This time factor defines an examination given under power or speed conditions. In a speeded test, score differences are determined by differences in the rate of response to
the test items; that is, the amount of time available is usually
limited, and those who respond at a slower rate may not finish
all of the items. The degree of speededness varies across tests.
The difference between a highly speeded test and a moderately
speeded test is the number of test takers expected to finish in
the time allowed. In contrast, in a power test the score differences are independent of the rate of response. That is, everyone
has enough time to respond to the items and relative scores
would not change if more time were available.

It is obvious that, from the test-taker's point of view, power
tests would be preferred. However, from the test-giver's point of
view, this is not always feasible or practical. It is also clear that
response rate is not always strongly related to accuracy of
response. Speed of response might be related to personality
characteristics like risk taking rather than to differences in
cognitive factors.

The results from the limited research available on the interaction of sex and speededness are mixed. Kappy (1980) looked
at the effects of speededness on the Graduate Record Exam
(GRE) for males and females. For both the quantitative and
verbal portions of the GRE, little evidence was found of differential speededness patterns for the sexes. Another study (Wild,
Durso, & Rubin, 1982) involving the GRE investigated whether
increasing the amount of time per question for the verbal and
quantitative sections of the exam would have a differential effect on examinee groups defined by sex, race, and number of
years since completing an undergraduate degree. The results
showed that although a larger portion of examinees were able
to finish the test when given additional time, this extra time did

not differentially improve the performance of any of the groups studied. In particular, females did not significantly increase their scores relative to the males.

Much of the research on test conditions investigates the effects of speeded tests on motivation and anxiety. In 1984, Hill, for example, examined the interaction of anxiety and time pressure and concluded that the low performance of high-anxious students is not due to lack of mastery of the material, but rather to motivational and test-taking problems that can be corrected. In a similar study, Plass and Hill (1986) looked at the interrelation of time pressure, test anxiety, and sex on third- and fourth-grade students taking a test composed of age-appropriate arithmetic problems. Using scores from a measure of test anxiety, the 173 students were divided into three groups based on low, medium, and high test anxiety. Approximately equal numbers of students from these groups were assigned at random by grade and sex into each of the two experimental testing conditions: one under time pressure and one in the absence of time pressure. Analyses of the data showed significant effects for the time pressure condition, level of anxiety, and sex. The children showed better performance without time pressure; low-anxious children scored higher than both middle- and high-anxious children, and females scored better than males. In addition there was a significant three-way interaction of time pressure, test anxiety, and sex. The authors report:

> In the condition removing time pressure, there are strong optimizing effects for boys but not for girls. Both high- and middle-anxious boys catch up completely with their low-anxious counterparts . . . In contrast, girls showed weaker interfering effects of anxiety, and there are no optimizing trends for high-anxious girls, who actually perform best under standard testing conditions. (p. 33)

The study also investigated the amount of time that students in the various anxiety groups took per problem. Using performance rate as the dependent variable in an analysis of covariance (ANCOVA), significant main effects for anxiety level and sex were found. High-anxious and low-anxious students took less time per problem than middle-anxious students and boys worked faster than girls. None of the interaction effects was significant. An examination of accuracy and performance rate revealed that high-anxious girls showed a slow rate with middle performance accuracy whereas high-anxious boys showed a fast rate with low performance. The authors summarized the importance of the rate-accuracy trade-offs in understanding performance and anxiety effects in testing as follows:

> The data indicate that there is an optimal, intermediate rate for high test performance, shown by low-anxious boys and girls in the present study. Middle-anxious children, especially girls, showed an accurate but too slow rate, whereas high-anxious children, especially boys, showed a too fast, inaccurate strategy. (p. 35)

The authors conclude that current testing programs could be improved if students were tested twice, once under standard conditions and once under optimizing (without time pressure) conditions.

Graf and Riddell (1972) evaluated the factor of time differently by measuring the effect of context on problem-solving performance and the amount of time used to solve the problems. Context was manipulated by presenting the subjects with two mathematically identical problems, one considered to have a context more familiar to females and one a context more familiar to males. Results showed that although males and females did not differ in the amount of time they took to solve the problem with the female context, females took significantly more time to solve the problem having a male context. The students' perception of the difficulty of the two problems was also measured. Males perceived that the two problems were of equal difficulty. The females perceived that the problem set in a male context was the more difficult. It is not clear whether this perception was a result of their experience with the problem (i.e., it took them longer to solve it and therefore they thought it more difficult), or whether they found it more difficult at the onset and therefore took more time in solving it. There was no difference in their accuracy in solving the two problems. The authors concluded that between-sex differences in problem solving could be significantly decreased by giving power tests rather than speed tests.

In summary, these studies do not strongly support the notion that time pressure differentially affects the performance of females and males on tests of mathematics. However, time pressure could interact with other individual variables, such as test anxiety, to result in differences for males and females (Plass & Hill, 1986). A conservative approach would suggest:

> that unspeeded tests of cognitive abilities should be used whenever sex-related differences are being investigated. It also suggests that using a speeded aptitude test as a criterion variable when examining sex-related differences in a typically male domain (e.g., mathematics) may be inappropriate or produce misleading data. In such male-typed areas, the true scores of high ability females may be underestimated. (Dwyer, 1979, p. 341)

ITEM DIFFICULTY SEQUENCING EFFECTS

Research on item sequencing effects investigates the differences in performance on achievement examinations as a result of changing the sequence of the test items. The most frequent arrangements are from easy-to-difficult, difficult-to-easy, spiral cyclical, and random. Arguments can be made for the potential merit of each of these arrangements. In the easy-to-hard arrangement, for example, beginning a test with easy questions could provide early success and therefore encourage continued effort. On the other hand, beginning a test with hard problems could challenge the student. In addition, the difficult items might be answered more easily when the examinee is less fatigued. Possible negative effects of this arrangement are also obvious. Examinees could become discouraged by encountering difficult items at the beginning of the test, especially if they thought the items would become increasingly difficult as they progressed. Spending time on the difficult items might result in not allowing enough time to answer the easy questions.

The results of studies on item arrangement that have included sex as a variable have been mixed. Plake et al. (1982) investigated the interactive effects on performance of the sex of the subject, test anxiety, item arrangement, and knowledge of arrangement on a mathematics test. The forty-eight-item multiple-choice mathematics test was composed of items from the ACT College Mathematics Placement program. It was considered slightly speeded. Three forms of the test were constructed using the item difficulty indices: easy-to-hard, uniform or spi-

ral cyclical, and random. For each form, half of the test booklets informed the examinee of the item arrangement and half
did not. Anxiety was a covariate. Results of the three-factor
fixed-effects ANCOVA (item order, knowledge of ordering, and
sex) showed a significant main effect for sex and a significant
sex-by-order interaction. Overall, males performed better than
females, and significantly better than females on the easy-to-
hard ordering of the items. Males also performed better than
females on the random item ordering. Knowledge of the item
arrangement did not appear to significantly influence test performance.

Plake, Patience, and Whitney (1988) investigated the effects
of item context on differential item performance between males
and females. The speeded test consisted of twenty mathematics
items selected by content from a pool of items. Three forms of
the test were assembled based on the difficulty indices: easy-
to-hard, easy-to-hard within content, and spiral cyclical. No
significant main effect was found for form or form-by-sex interaction. A significant main effect was found for sex, with females
outperforming the males. Plake concluded "that item arrangement is not a potent variable in producing differential item
performance between males and females" (p. 892).

Similar results were found in a study (Klimko, 1984) involving college students in an introductory educational psychology
course; it examined the effects on test performance of item
arrangement, cognitive entry characteristics, test anxiety, and
sex. Three forms of the fifty-item multiple-choice midterm examination were used: easy-to-hard, hard-to-easy, and random.
Of the four independent variables, the only significant predictor
of achievement differences was student cognitive entry characteristics. Item arrangements based on item difficulties and sex
did not influence performance. The author cautioned against
drawing conclusions based upon gender due to the small number of males in the study.

Item order and sex were among the variables considered in
a study of fourth graders by Kleinke (1980). Two forms of the
speeded social studies test were used: easy-to-hard and uniform. Although the boys outperformed the girls, there were no
significant interaction effects for sex and item ordering. There
was a significant effect for item-ordering with those examinees
taking the easy-to-hard form; they scored higher. The author

concluded that item order should be considered on speeded tests.

One variation on item arrangement studies is student awareness of the item arrangement. In a two-experiment study Lane et al. (1987) developed five forms of a forty-item multiple-choice exam. The items tested course content from an undergraduate education course. Item order was determined as a result of manipulation of the statistical and cognitive item-difficulty level (based on Bloom's taxonomy). In the first experiment, students were unaware of the ordering patterns. The results showed no significant differences for the test of item order by gender. In the second experiment, six forms of the test were developed by manipulating the statistical and cognitive item-difficulty level. Knowledge of the ordering was provided by labels that indicated the cognitive level of the item. The results showed no significant differences for item order by gender by knowledge. However, a significant difference was found for the interaction of knowledge of level with gender. Males without labels scored lowest, followed by females without labels, then by males with labels. Finally, females with labels scored the highest. The authors concluded that the lack of an item ordering by gender interaction in the second experiment suggests that the long accepted view that easy items should come first is oversimplified. They offered no conclusions based upon gender except to note the increase in the males' scores when labels were provided.

Hambleton and Traub (1974) also investigated the effects of item arrangement for males and females, using a mathematics achievement test of multiple-choice items arranged easy-to-difficult and difficult-to-easy. Since the amount of time was restricted to forty minutes, the test was considered slightly speeded, although the differences in the number of students completing each form was not significant. Neither the main effect due to sex nor the interaction between item order and sex was significant. They did find, however, a significant main effect due to item ordering with higher mean scores on the easy-to-difficult arrangement. The authors concluded that reordering the items on a test produces a test with properties different from the original.

In a review of these and other studies that considered item arrangement but not gender, Leary and Dorans (1985) concluded that hard-to-easy arrangements of items should be avoided for all students, especially under highly speeded conditions. No additional conclusions regarding the differential effects of item arrangement due to the gender of the examinee seems warranted, based on the studies reviewed here.

EXAMINATION FORMAT

Examination format is another factor in studies of differential performance for males and females. The two formats usually considered are essay and multiple choice. The arguments for the importance of format are: superior verbal ability in females will enhance their scores on essay exams, and differences in test-taking strategies will favor males on multiple-choice exams. However, format as a factor is perhaps not as relevant for mathematics tests where alternatives to the multiple-choice format would not place much emphasis on writing ability. As with other test factors reviewed above, not all of the studies reviewed used mathematics as the content of the test.

Murphy (1982) studied the performance of male and female candidates on sixteen General Certificate of Education (GCE) examinations. The examinations included both multiple choice and other forms, and three of the sixteen tested mathematics achievement. At least one thousand males and one thousand female students took these tests for each of four consecutive years. The overall performance of females relative to males was not of interest in this study, but rather the relative performance of the groups on one type of exam compared to the other types. A series of t-tests was carried out to determine any significant difference in the performance of male and female examinees on the multiple-choice tests, as compared to the other formats. Results showed that in the majority of cases the males performed better than the females on the multiple-choice exams, as compared to their relative performance on the other formats. An important exception occurred on two of the three mathematics tests. On these two tests, there was no consistent male advantage relative to the performance of the females on the multiple-choice tests. The fact that a male advantage did exist on the one exam is an unexplained inconsistency.

Using a sample of fifteen- and sixteen-year-old Irish students, Bolger (1984) examined gender differences in achievement for three school subjects (Irish, English, and mathematics). Multiple-choice and written formats were used. Males were found to perform relatively better than females on the multiple-choice forms and relatively poorer on the written examinations; the opposite was true for females. The effect was constant across the three subjects measured. An additional hypothesis that gender difference would be largest for the languages and smallest for mathematics was not supported. The author cited this as evidence that method-based gender difference cannot be attributed to the differential verbal skills required by the two methods. Alternative explanations offered by the author include neatness of presentation contributing to the performance of females on essay exams and the possibility that males are more likely to guess the answer on multiple-choice exams and, therefore, be more likely to obtain the right answer.

Studies testing for gender differences on multiple-choice exams do not always reveal a male advantage. In a test of English language comprehension and composition, Bell and (Hay, 1987) used both multiple-choice and extended-response formats. Males were not found to score higher on multiple-choice questions.

These three studies examined the relative difference in scores between multiple-choice and essay format exams for males and females. Gender differences did not always occur, but when differences were found, they favored males on multiple-choice exams. It is reasonable to conclude that exam format does contribute to gender differences and that the use of the multiple-choice format can result in a male advantage that is independent of ability.

Student behaviors associated with multiple-choice exams (e.g., guessing and answer changing) have also been studied. Differences in these behaviors for males and females might explain the differences found overall on multiple-choice exams. For example, if females are more likely to answer only those questions on which they are sure of the answer and leave the rest blank, they will score lower than equal ability males who respond to the questions they know and guess on the questions for which they are unsure.

Choppin (1975) looked for gender differences in the tendency to guess on multiple-choice exams. The sample consisted of fourteen-year-old students from 160 secondary schools in

England and Wales who had participated in the cross-cultural International Association for the Evaluation of Educational Achievement (IEA) study of academic standards. The study analyzed responses to six separate multiple-choice instruments that tested aspects of science and the English language. The results showed a significant difference in favor of males in the tendency to guess. However, the size of the gender difference was small compared to the size of the difference found when the data were analyzed by school type. A similar battery of tests was administered to ten-year-old students and again the data were analyzed for gender differences. For this sample no clear differences emerged, although, for both males and females, the tendency to guess remained high.

More recently Khampalikit (1982) investigated guessing as a test-taking strategy of elementary school students. Random samples of students in Grades 2, 5, and 8 from a nationwide norming group were used to compute four guessing-related indices using item responses for the 3R's Reading Test and Mathematics Test. Results showed that the overall amount of guessing was low and there was little evidence of differences between the sexes on the test-taking behaviors assessed.

Answer-changing behavior was the subject of a study by Skinner (1983). Males and females from an introductory psychology course served as subjects. The test was a speeded 100-question multiple-choice exam given as a midterm examination. Erasures on the answer sheet were examined to determine answer changes and whether the answer was changed from right to wrong or wrong to right. Overall, the number of answer changes was small, only about 4 percent. An analysis for gender differences revealed that females made significantly more changes than did males. Their rate of 4.83 percent was more than double that of the males (2.36 percent). Regarding this gender difference, the author concludes:

> Clearly, deliberating about answer-changing leaves less time available for other activities, such as answering multiple choice questions not yet attempted, or doing other types of questions (e.g., essays). Thus, regardless of whether or not there is a functional relationship between the number of answer changes and time taken to

consider and implement such changes, *on a speeded test* the tendency for females to make more than twice as many answer changes as males may well be counter-productive, particularly in light of two further findings: first, the success rate for answer changes for women was not better than that for men (indeed, males made 54% successful changes, females 50%); and second, female subjects achieved a mean grade of 65.9% on the examination, compared to 70% for the males. (p. 221)

Answer-changing behavior has been studied in relation to other test-taking factors, such as, for example, test anxiety. Payne (1984) hypothesized that high test anxiety would be associated with a high degree of answer changing. Data for a sample of 296 eighth-grade students consisted of scores on an anxiety measure and the number of item revisions on an aggregate of four multiple-choice science achievement tests given over the period of one year. Item changes were coded wrong-to-wrong, wrong-to-right, and right-to-wrong. Race (black and white) and sex were also used as factors in the study. No significant sex differences in answer changing were found. Significant correlations between answer changing and test anxiety were found for white males, for the total white student group, for the total male group, and for the total student group, suggesting a tendency in each for higher test anxiety to be associated with more answer-changing behavior. None of the correlations for black students was significantly different from zero.

Answer changing and guessing might be associated with gender differences found on the 1988 University of Minnesota Talented Youth in Mathematics Program (UMTYMP) testing for the Twin Cities sample (Terwilliger, 1988). Response patterns for the 50-item multiple-choice test were analyzed on a gender basis. It was found that females were less likely than males to finish the test, and the drop in their success rate toward the end of the test was more pronounced than that of the males. The responses were not analyzed for answer changes.

Although the evidence is not conclusive, these studies suggest that guessing and answer changing might be associated with gender differences on multiple-choice exams. A greater tendency by males to guess answers can result in higher test scores independent of ability. Likewise, a greater tendency by females to change answers can result in lower test scores because of the time this behavior takes in a speeded test.

TEST-WISENESS AND RISK TAKING

Two additional factors that are sometimes associated with differences in test taking are test-wiseness and risk taking. Test-wiseness (TW) has been defined as "a subject's capacity to utilize the characteristics and formats of the test and/or the test taking situation to receive a high score" (Millman, Bishop, & Ebel, 1965, p. 707). Risk taking on objective examinations (RTOOE) is defined as "guessing when the examinee is aware that there is a penalty for incorrect responses" (Slakter, 1967, p. 33). Both TW and RTOOE are associated with the multiple-choice format and, therefore, gender differences in them might help explain gender differences due to test format.

Slakter, Koehler, and Hampton (1970) developed a measure of test-wiseness for use with students in Grades 5 through 11. The instrument measured four aspects of TW corresponding to these behaviors: (1) select the option which resembles an aspect of the stem; (2) eliminate options which are known to be incorrect and choose from among the remaining options; (3) eliminate similar options, that is, options that imply the correctness of each other; (4) eliminate those options which include specific determiners (p. 119). The instrument was administered to 1,070 students in Grades 5 through 11 and replicated with a group of 1,291 students. A sex-by-grade multivariate analysis of variance was performed on the four subscale scores. The only significant effect was that of grade with TW increasing over grade level. Neither the sex effect nor the sex-by-grade interaction was significant.

A related study (Crehan et al., 1978) was conducted to determine the relationship between TW and grade level, the relationship between TW and sex, and the stability of TW. This was a longitudinal study that tested students three times at two-year intervals in Grades 5, 7, and 9 ($n = 75$); 6, 8, 10 ($n = 76$); 7, 9, 11 ($n = 73$); and 8, 10, 12 ($n = 64$). The same four aspects of TW were measured. As with the previous study, there was no evidence of sex differences or sex-by-grade interaction. TW was found to be somewhat stable and to increase over grades.

RTOOE has been measured using a variety of formulas. Slakter (1967) compared results obtained using five different measures of RTOOE and an additional one he proposed in this study. Results did not show any sex differences in RTOOE for any of the measures.

In summary, there is no evidence that males and females differ in terms of test-wiseness and risk taking on objective examinations.

TEST PREPARATION STRATEGIES

The final area to be examined is that of study behaviors and test preparation strategies for males and females. This research has focused on characterizing the study behaviors of successful students. One general conclusion is that no single approach is associated specifically with success. Biggs (1976) looked for differences in study behaviors for males and females. Using the Study Behavior Questionnaire developed for his study, he found evidence to suggest that a single task can be successfully approached in different ways by males and females. He characterized the male approach as "seeing 'truth' emerging from external sources and authorities, and not worrying too much about interrelating past knowledge with what one is in the process of acquiring" (p. 77). On the other hand, he described the approach taken by females to be that of "making up one's own mind about 'truth' by avoiding rote learning of detail and by actively using transformational strategies" (p. 77). It is interesting to note that these characterizations of study behavior are exactly opposite those that are usually put forth to explain superior male performance on standardized mathematics achievement tests.

Watkins and Hattie (1981), also using the Biggs Study Behavior Questionnaire, investigated the study methods of students at an Australian university. Other factors that they considered in addition to gender were age, academic year, and field of study. They found that regardless of these other factors,

> females were more likely than the males to show interest in their courses and to adopt a deep-level approach to their work. At the same time the females also generally seemed to possess more organized study methods

than the males. The males were more likely to have a pragmatic approach to tertiary study, to be more worried about their work, and to adopt reproducing strategies which would allow them to scrape through their examinations. (p. 392)

They further noted that, based upon these findings, females would be expected to achieve better academic results than males; indeed, this proved true for the females they investigated.

These results are interesting when applied to gender differences in mathematics achievement. In a review of the literature, Kimball (1989) pointed out that in contrast to standardized measures of mathematics achievement, females receive better grades in school mathematics than do boys. If these same differences in academic behaviors, as observed by Watkins and Hattie (1981), are present at the elementary and secondary school levels, they might help explain the higher grades earned by girls. They do not, however, explain why boys score better on standardized tests.

A study by Speth (1987) investigated the interaction of learning style, gender, and type of examination on anticipated test preparation strategies. The two examination conditions used were multiple choice and essay exams. On the basis of two different learning style instruments, the students from educational psychology classes were grouped into four clusters. A survey of test preparation activities developed by the investigator was used as the dependent measure. A factor analysis of the test preparation survey yielded six subscales. A 4 × 2 × 2 MANCOVA tested hypotheses of no difference among clusters, between males and females, or between test conditions on the six test preparation strategies, while controlling for a self-rating of academic ability. The results showed no significant main effect for sex or a sex-by-test type interaction. There was, however, a significant three-way interaction of cluster, gender, and type of examination. This suggests that gender by itself is not the critical variable, and that males and females relative to each other do not prepare differently based on the method of testing. Instead, as Biggs found, different test preparation strategies correlate with gender and learning styles (Biggs, 1976). What was not included in this study was any measure of the success of these different test preparation strategies.

This report presents an overview and conclusions from three studies, *The IMPACT Project* (Clarke, 1985), *Assessment Alternatives in Mathematics* (Clarke, 1989), and the *Vaucluse College Study*, which suggest how the above standard can be realized at classroom and school level. The central purpose of this chapter is to discuss:

- the extent to which the strategies reported encourage children to broaden their mathematical thinking and facilitate metalearning.

- the impact of these strategies on the nature of mathematical activity in classrooms and, in particular, with regard to redefining the roles of teacher and student in creating and giving personal meaning to mathematics.

The nature of these strategies was essentially metacognitive. The meaning of metacognition in the mathematics context has been most usefully articulated by Garofalo and Lester (1984) as the knowledge and regulation of cognition. An essential aspect of this metacognitive activity is reflection on learning (Kilpatrick, 1985), and White (1986) identified a need for training in just this aspect of metacognition: "Much learning is superficial, being done without deep reflection. Appreciation of this point leads to the recognition of the need for training in metacognition" (p. 5).

Biggs (1988) put forward the image of "learning through guided student self-questioning" and suggested "self-management of learning" as an essential goal of education. The reflective review of learning and student self-management of learning were central concerns of all three studies reported here.

The learning of mathematics is fundamentally a matter of constructing mathematical meaning. The environment of the mathematics classroom provides experiences which stimulate this process of construction. While the mathematical knowledge of school children will incorporate visual imagery, both at the level of ikonic thought and at a level involving more elaborated visual representations (geometrical, graphical), mathematical meaning requires a language for its internalization within the learner's cognitive framework and for its articulation in the learner's interactions with others. Communication is at the heart of classroom experiences which stimulate learning. Class-room environments that place particular communication de-mands on students can facilitate the construction and sharing of mathematical meaning and promote student reflection on the nature of the mathematical meanings they are required to communicate.

FACILITATING COMMUNICATION IN THE MATHEMATICS CLASSROOM: THREE STUDIES

We would suggest the existence of three distinct types of com-munication in the mathematics classroom:

- communication about mathematics;

- communicating mathematics;

- using mathematics to communicate.

The IMPACT Project is primarily concerned with the first of these. The use of student mathematics journals at Vaucluse College provided the opportunity for the development of com-munication in all three modes. The IMPACT program required that students reflect on their mathematical activity and their learning and, through student-teacher dialogue, sought to fa-cilitate self-management of learning. The national development and testing of the assessment strategies which became Assess-ment Alternatives in Mathematics (Clarke, 1989) gave consider-ation to communication in two senses: communication as the means by which assessment information is obtained and com-munication skills as one focus of assessment. The use of math-ematics journals in the Vaucluse College study demonstrated the potential to develop in learners the reviewing and reflective skills required by the IMPACT program and also to develop in students the ability to think mathematically.

THE IMPACT PROJECT

The IMPACT Project benefited from the support of the Faculty of Education, Monash University. The publication of the evalu-ation report (Clarke, 1985) was funded by the Monash Math-ematics Education Centre. The objectives of the IMPACT project were:

- to provide a mechanism whereby the student can regularly inform the teacher of difficulties experienced, help needed, and anxiety felt;

- to encourage and facilitate meaningful pupil-teacher dialogue, student reflection on learning, and negotiated instruction.

During 1984, about seven hundred Year 7 children in thirty-six mathematics classes in fifteen Victorian secondary schools were regularly given the opportunity, about once every two weeks, to give confidential, written answers to questions like:

What was the best thing to happen in Maths?

What is the biggest worry affecting your work in Maths?

What is the most important thing you have learned in Maths?

How do you feel in Maths classes?

How could we improve Maths classes?

The regular, written reflections of seven hundred children concerning their learning of mathematics provided a pool of data related to the achievement of the above objectives. A graphic portrayal of the children's conception of mathematics emerged over the year (Clarke, 1987). The immediacy of this portrayal was heightened by the children's spontaneous (and highly idiosyncratic) use of technical mathematical terms. Many of the current preoccupations of the mathematics education community (language, differential treatment and behavior of boys and girls, mode of instruction, student-generated algorithms, the social context of instruction and learning, and so on) emerged as concerns for the students in this study. Participant teachers expressed surprise at the significance of these issues for their classrooms.

Examples of questions used and some answers obtained follow:

What would you most like more help with?

Nothing much, but I'm not sure	Your way	4⟌80̅
how to do division your way.	My way	4⟌80̲

Write down one particular problem which you found difficult.

Algebra a bit, because I don't understand why we don't just use numbers. It would be simpler.

Write down one new problem which you can now do.

$^1/_3 \div 4 = ^1/_3 \times ^4/_1 = ^4/_3 = ^1/_{12}$

How could we improve maths classes?

Have less work and more learning.

The nature of communication in the mathematics classroom became the central consideration in the teachers' evaluations of both instruction and learning and in their planning for further instruction. The IMPACT study provided an opportunity with some classes for a redefinition of the function of the mathematics classroom. The findings show clear instances where teacher action in response to student requests or suggestions significantly altered the form of instruction. Students in those classes were confronted with the need for a reinterpretation of their role in determining the nature of classroom activity and the possible nature of student-teacher communication.

Specific findings from this study included:

- The student attitude to the administration of the IM-PACT procedure was predominantly one of acceptance and passive compliance.

- The quality of student responses varied. Teachers reported that many students experienced difficulty in articulating their feelings or their mathematics difficulties.

- A majority of students reported finding the procedure "useful."

- More boys than girls reported finding the procedure personally useful. (An interesting result, since several teachers commented that the girls made better use of the procedure, offering more informative and insightful responses.)

- Students who found the procedure useful offered three categories of benefit:

Reflection on Learning—"It makes you realize more about the subject" and "It makes you think how you're going."

Reporting Feelings—"It gives you a chance to express your feelings/tell about your problems/say what you like and don't like."

Information for the Teacher—"It helps the teacher know . . . "

• Teacher action took the following forms: Organizational action, instructional action, individual assistance, individual counseling and, in two cases, no action.

• Most teachers reported improved student-teacher relationships

• Where a student expressed dissatisfaction with the procedure, the reason most commonly given was lack of consequent teacher action.

• Several instances were documented in which teacher action arising from information provided through the IMPACT procedure led to positive changes in student attitudes and achievement.

• Over 80 percent of participating teachers consistently reported finding the IMPACT procedure to be of value.

Teachers identified a lack of time in the past to engage in private conversation with every pupil as a major concern. As a result, they greeted the IMPACT procedure with initial enthusiasm, since it provided the opportunity for all students to communicate confidentially with their mathematics teachers with minimal reduction in instruction time. Other benefits were identified by the teachers:

Students talked to me through the sheets, very frankly, and I gained tremendous insights into anxieties they had, and frustrations. Students I felt were coping quite happily mentioned anxiety about tests. Some students felt I did not explain things thoroughly enough and went too fast, and these were students who did quite well in tests, so I had assumed they were happy. Other students mentioned boredom and felt the work was too easy and was repeating Primary School. Without the sheets, I would not have gained this information as they would never have been so frank in conversation. I obviously changed my teaching methods to comply with the information and this helped my relationship with the class and with individual students. In general, students really appreciated the fact that I was taking the trouble to find out what they think and they used the system very responsibly.

(Year 7 Maths teacher, female, September 1985).

Use of the IMPACT program facilitated communication between teachers and students about the mathematics being studied and about the students' feelings concerning their learning, the content, and the instruction. In several instances, this communication led to fundamental changes in instructional practice, learning behavior, and classroom environment.

The extent to which participation in the IMPACT program actually *facilitated* metalearning and the development of student mathematical thinking remains uncertain. The IMPACT program certainly provided a stimulus for reflection on learning, but no training was provided in review techniques or metacognitive strategies (cf. the PEEL Project, Baird & Mitchell, 1986). Nor was any feedback provided to students concerning the quality of their IMPACT responses. By its nature, the IMPACT program provided documentation of student communication of the first type, communicating about mathematics, and, to a lesser extent, of the second type, communicating mathematics.

ASSESSMENT ALTERNATIVES IN MATHEMATICS

In 1986, the Mathematics Curriculum and Teaching Program (MCTP), a national initiative concerned with the professional development of mathematics teachers, commissioned a study of effective assessment practices in use in Australian mathematics classrooms. The outcome of this project was to be a teacher resource guide, subsequently published as *Assessment Alternatives in Mathematics* (Clarke, 1989). It was aimed at assisting mathematics teachers to expand their repertoire of assessment strategies in order that their assessment might become optimally effective, giving appropriate recognition to all the goals of the contemporary mathematics classroom.

Communication in the mathematics classroom became a central concern in the compilation of assessment strategies. This was particularly the case with regard to the assessment of student problem solving in mathematics, where it became evident that a teacher's capacity to evaluate a student's problem-solving performance was critically dependent on the student's ability to articulate, in either spoken or written form, the problem-solving process, the nature of the solutions, and the evaluation of the appropriateness and the quality of their solutions.

Among the various assessment strategies collected, studied, tested, and refined during the course of this project, the role of communication varied with the particular strategy under consideration. The following discussion examines the nature of the communication component for a sample of the assessment strategies.

ASSESSMENT THROUGH CLASSROOM OBSERVATION

Teachers made succinct annotations to class lists during the course of a lesson. These brief records were restricted to "aberrations and insights," that is, observations of student behaviors or utterances which challenged or extended the teacher's existing conception of a student's competence or understanding. The effectiveness of such informal assessment is critically dependent on the nature of classroom activity. This was addressed directly by drawing teachers' attention to factors which facilitate or inhibit student communication in the classroom. A major assertion of Assessment Alternatives in Mathematics was that the most effective instructional activities are typically those which also provide the best assessment opportunities. Strategies for maximizing assessment opportunities included consideration of "wait time" (Rowe, 1978), the characteristics of "good questions," and the establishment of "student work folios."

Test Alternatives

Teachers were encouraged to explore different approaches to formal testing. These included:

Practical tests, in which student competence was demonstrated through the completion of tasks with a practical emphasis, typically involving the manipulation of concrete materials. Computing skills were also assessed in this way. It was a common requirement for students to provide an account of their methods, but the essence of this approach was communication by demonstration.

Group tests, in which tasks were solved through student collaboration. Successful performance was associated with effective student-student communication and an ability to translate into personal terms the ideas and insights of others.

Student-constructed tests, in which groups of students would contribute test items covering a topic just completed. Trial teachers were unanimous that the demands of articulating the essence of a topic through a representative set of problems made this strategy an immensely powerful review technique, leading to significant advances in student understanding. The resulting tests were consistently more difficult than those the teacher would have set, were typically completed with higher levels of student enthusiasm and success, and provided a context particularly conducive to subsequent discussion.

Problem Solving and Investigations

A four-dimensional structure for the assessment of problem-solving behavior emerged in the course of the testing (see, for comparison, Schoenfeld, 1985), and teacher attention was drawn to the need to identify which aspect of problem-solving behavior was of interest. Assessment information was typically collected through informal observations and from student reports. This information could then be located within the categorization scheme below.

Dimension 1 relates to the spontaneous use of mathematical procedures, principles, and facts, that is, the mathematics that our students *choose to use*—without the explicit cueing of a test question.

Dimension 2 is concerned with problem-solving strategies. There are many lists of such strategies. Practicing teachers seemed quite confident in their ability to distinguish strategies such as "restated the problem," or "organized information systematically," or "found a related but similar problem" (and so on) from the mathematical tool skills which provide the focus of Dimension 1.

Dimension 3 is the structural dimension, particularly concerned with planning, decision making, verifying, and evaluating. One secondary mathematics teacher provided a succinct summation of the focus of Dimension 3 in observing, "Students should show a systematic approach of reviewing what they know, planning their actions, testing their ideas, and evaluating their work."

Dimension 4 is the personal dimension, concerned with student participation, motivation, work habits, the skills associated with cooperative group work, and beliefs about the nature and purpose of mathematical activity.

Teachers reported that students experienced significant difficulties in recording and reporting their problem-solving attempts and required substantial guidance and detailed feedback.

Communicating Assessment Information

Issues related to the grading of student work and the effective reporting of assessment information were explored. The need for clarity of communication and the establishment of an ongoing dialogue between student and teacher concerning the student's growth towards competence was stressed.

Expanding the Assessment Network

Teachers were encouraged to consider other purposes to which assessment information might be put (program evaluation and instructional review, for instance) and other groups or individuals who might contribute assessment information. Parental involvement, peer tutoring, and peer assessment were investigated and various strategies offered to facilitate student self-assessment. These latter strategies included the IMPACT procedure, already reported, and the use of student mathematics journals. It was the evaluation of the use of student journals which subsequently became the focus of the Vaucluse Study reported below.

Communication plays a central role in each of these approaches to assessment, and much of the effort expended during the testing of the assessment strategies related to fostering clear, purposeful, meaningful, informative communication of mathematics and about mathematics.

THE VAUCLUSE COLLEGE STUDY

This study explored the implications of the regular completion of student journals in mathematics. Vaucluse College is a Catholic secondary girls school. There are approximately five hundred girls from Year 7 to Year 12 at Vaucluse. It serves a multicultural population: 20 percent Asian, 30 percent Italian and Greek, with the remaining 50 percent being predominantly Anglo-Saxon. For all students at this secondary school from Year 7 onwards, a central component of mathematical activity is the daily completion at home of a student journal. Through their journal-keeping activities students are introduced to describing what they have learned, summarizing key topics, and identifying appropriate examples and questions. Regular monitoring of the journals informs teaching practice and provides the basis for individual teacher-student discussion.

In 1986, mathematics journals were first introduced experimentally in one class each at Year 7, 9, and 10 levels. Results were encouraging enough to warrant the expansion of their use. By the start of 1989, the keeping of mathematics journals was seen as an essential element in the teaching of mathematics from Years 7 to 10. Appendix E presents the history and rationale for student mathematics journals at Vaucluse College from the perspective of the school and the mathematics staff. The school statement includes the following aims:

By keeping a mathematics journal we intend that students:
1. Formulate, clarify, and relate concepts.
2. Appreciate how mathematics speaks about the world.
3. Think mathematically:
   a. Practice the processes (problem solving) that underlie the doing of mathematics.
   b. Formulate physical relations mathematically.

As an introduction to journal writing, Year 7 students are supplied with a book in which each page is divided into three sections: What we did, What I learned, Examples and questions. Students are required to write in their journal after every mathematics lesson. This is seen as ongoing homework. Journals contribute 30 percent to the assessment in mathematics. When writing student reports, mathematics teachers were given the following guidelines for the assessment of students' journals:

A. Quantity of work.
   1. Frequency: that is, is it done after every lesson?
   2. Volume: the amount of work done can be taken as a measure of both ability and enthusiasm.
   3. Presentation.

B. How well is it used?
   1. Is the work summarized, and do the summaries indicate developing note-taking skills?
   2. Is the journal used to collect important examples of procedures and/or applications?
   3. Are errors identified and discussed?
   4. Are there signs of involvement with the work, original or probing questions, a willingness to explore?
   5. Is the student learning to "dialogue," that is, ask her own questions and then set about methodically seeking an answer and presenting her investigations logically?

As a minimum, a satisfactory journal entry should reflect the intellectual involvement of the student in the day's lesson. What form a particular entry will take is determined by the form of the day's lesson and the level of sophistication at which the student can interpret the journal tasks. Appendix E sets out the school's expectations with regard to theory, practice, and activity-oriented lessons.

Journal writing was intended to assist students to see themselves as active agents in the construction of mathematical knowledge (see Stephens, 1982). The school hoped that journal writing would assist students progressively to engage in an internal dialogue through which they reflected on and explored the mathematics they met. In this respect, there is a link to the IMPACT Study through a similar focus on the development of metacognitive learning. It was also hoped that students, through their journal writing, would begin to see mathematical activity not simply in terms of applying prescribed rules and procedures, but more as engaging in activities such as searching for patterns, making and testing conjectures, generalizing, asking "Why?", trying to be systematic, classifying, transforming, searching for methods, deciding on rules, defining, agreeing on equivalences, reasoning, demonstrating, expressing doubt, and proving (cf. Mason, 1984).

If such aims were to be realized through the use of journals, it would be necessary to focus on the linguistic forms by which students communicated what they had learned and how they had gone about it.

Methodology of the Evaluation Study

During 1988 and 1989, an evaluation was conducted of student journal use and its effects on the learning and teaching of mathematics. Consultation with school staff and perusal of a sample of student journals led to the construction of a questionnaire which, after testing, was administered to all students in Years 7 to 12. The questionnaire examined student use of journals and their perceptions of the purpose of journal communication and its contribution to their learning of mathematics. Students' conceptions of the nature of mathematics and of mathematical activity in schools were also addressed. A similar survey was conducted of school mathematics staff, with specific focus on the extent to which they valued and fostered students' journal communications and made use of student journal communications in their classroom teaching and in their work with individual students.

At the time the evaluation began there was a perception in the school mathematics department that a progression existed in student journal writing from a narrative mode to a summary mode to dialogue. Conversation with teachers and the perusal of student journals suggested that student journal writing could be usefully divided into the three categories: Narrative (or Episodic), Summary, and Dialogue. This categorization assumed the status of a hypothesis, and provided much of the structure for the initial data analysis. School sources asserted that a major aim of journal writing was to facilitate student development in question asking and that questioning reflects the dialectic of Narrative, Summary, and Dialogue (Waywood, 1988).

Narrative, Summary, and Dialogue

The categorization of student journal use into Narrative, Summary, and Dialogue warrants more detailed explanation. The examples which follow were offered as both illustrative instances of each response category (Examples 1, 4, and 6), and also as examples of "transition" responses from students whose journal entries suggest that they are in transition between categories. Seen in this light, Example 2 shows a student moving from simple narrative of classroom experiences to the restructuring of content and experience required for effective summary. Examples 3 and 5 show two students' initial experiments with a new form of journal entry. In each of these two cases, the excerpts represent embryonic instances of the Summary and Dialogue categories, respectively.

Narrative:
Example 1. "Today was the day that Mr. Waywood was absent and set us work to do that gave me a lot of thinking to do. I don't think that it was very hard but you had to think about what to write for the answer to the questions."
Example 2. "I think today I began to understand that maths is a way of describing things in reality. A great example is that a ball flying through the air travels the path of a parabola. Because there is an infinite number of ways for the ball to travel there is an infinite number of possible parabolas. Because parabolas can be written mathematically there would be a mathematical function to describe every arc in the world."

Summary:
Example 3. "Logarithms are an index which are used to simplify calculations. The whole number part of a logarithm is called the characteristic. The decimal part of a logarithm is called the mantissa."
Example 4. "Equations . . . the main word here is to solve. Equations have an unknown—there is an answer to the problem. Linear techniques revolve around inverse operations, and quadratic equations, different from the above, require different techniques to solve them, such as factorization. . . . You can't solve all the equations the

same way, because they are all different, and that is why we have to learn different techniques."

Dialogue:

Example 5. "The sin of 60 = 0.866025403 . . . firstly is the sin of 60 infinite I wonder. I think it is because you said the points on a circle are infinite. Then how could the square of 0.866025403 . . . be exactly 0.75? If it is just an approximation, then how could it equal exactly 1? Can you please explain?"

Example 6. "Another thing, transposition and substitution, really show you the quality of operations. Like division, is sort of a secondary operation, with multiplication being the real basis behind it. This ties in with my learning about reading division properly (in previous pages), that is, fractions are different forms of multiplication. So I guess that's like rational numbers (Q) are like a front for multiplication, an extension of multiplication. Which came first, multiplication or division? It would have to be multiplication. They are so similar, no that's not what I mean. I mean they are so strongly connected. But its like division does not really exist, multiplication is more real. The same with subtraction. Addition and Multiplication are the only real operations."

The study design provided a diversity of data sources by which the validity of the categorization could be assessed. Student interviews, student and teacher questionnaires, teacher interviews, and the study of journal entries represented a substantial body of data by which both the individual validity of each category could be judged and any patterns of individual development identified.

Observations and Findings

An initial analysis of the student survey data has been completed. Findings suggest that journal writing leads to a progressive refinement of purpose from an initial narrative stage of simply listing events in the mathematics classroom to summarizing work done and topics covered. Within this stage, we note a move away from a simple summary of items of mathematical work covered to a more personal summary of mathematical activity in terms of developing understanding and addressing problems. Finally, some students move beyond this to an internal dialogue, where they begin to pose questions and hypotheses concerning the mathematics in which they are engaged (e.g., "I wonder whether this works for other graphs as well," and "So, why is it that . . . ").

More specifically, the narrative descriptions of what was done on a particular day, so prevalent in Years 7 and 8, appear to be progressively enriched by the inclusion of reflective writing in which the students discuss how they went about an investigation and how the work in hand related to work they had previously covered. This review process, together with requests for teacher help and indications of things they would like to find out, is similar to the responses solicited through the IMPACT program. Journal entries of some students occasionally took on the aspect of dialogue. Our research suggests that through the process of their journal writing students increasingly interpret mathematics in personal terms, constructing meanings and connections.

Student Survey Findings

While questionnaires were administered to every student, a sample of 150 students, 25 at each year level, was chosen for statistical analysis. Three questionnaires were administered ("Mathematics," "Journals—Part A," and "Journals—Part B," in that order), and the sample selection procedure ensured that all students at a particular year level, who had completed all three questionnaires, had the same chance of appearing in the sample.

A full statistical report was prepared for the use of the school (Clarke, Stephens, & Waywood, 1989), but the purposes of this report are best served by a summary of significant findings. These are set out below, with related conclusions appropriately clustered. It must be borne in mind that these findings are the results of students' reports of their behavior, their teachers' behavior, their perceptions, and their beliefs.

Frequency of Journal Use.

- The majority of students (54 percent) reported that they write in their maths journal "after every lesson."

- A similar majority (53 percent) estimated the time spent on journal writing in one week as less than one hour.

• Ninety percent of students reported reading their journals either occasionally or often.

Nature of Journal Use. By clustering student responses to particular items it was possible to construct indices associated with the hypothesized taxonomy of writing modes: Narrative, Summary, and Dialogue. Of the sample of 150 students, 65 could be identified as predominantly employing one of the three modes of journal use. This enabled statistical analyses to be carried out for this subset of students incorporating a measure of Mode of Use. (A "Modal Rating" on a seven-point scale was subsequently generated for 123 of the 150 students, and the conclusions which follow held true for both measures).

• Year Level was more decisive in determining the frequency of journal use than was a student's experience with journal use. However, experience with journal use was more significant in accounting for Mode of Use. This justifies the conclusion that *it is the experience of using journals that promotes more sophisticated modes of use rather than simply student maturation.*

• Analysis of variance revealed that Mode of Use made the most significant contribution to accounting for the variation evident in the three key indices, User Index, Difficulty Index, and Positive-Effect Index.

• A clear and statistically significant trend emerged in the consideration of Mode of Use in relation to each of the other critical indices. *The more sophisticated the mode of journal use, the more likely a student was to:*

    * make more use of journals
    * find journal completion less difficult
    * express greater appreciation of journal completion
    * report positive, rather than negative, outcomes of journal use.

These results may not be surprising, but the consistency in the direction of the trend and in the statistical significance strongly supports the interpretation of Mode of Use as a meaningful structure for the analysis of student journal writing.

Incentives and Obstacles to Journal Use

- Sixty percent of students gave as the main reason for writing in their journal, "because it helps me." In another item, the most popular justification for journal use was "to help me learn."

- Most students (75 percent) found the act of journal writing "mostly" or "always" easy. However, students were evenly divided over whether or not they found it difficult to put their mathematical thinking into words. In this regard, it is worth noting that half of the student sample reported that the most important thing learned from journal completion was "To be able to explain what I think."

Purpose.

- Asked to identify "the most important thing for me to do in my journal," students indicated: "to summarize what we did in class," "to write down what I understand," and "to write down examples of how things are done," in that order. These responses are consistent with the finding that the majority of students appear to be operating in the Summary mode and to perceive journal use in either Summary or Narrative terms.

- In response to the item "I think of my mathematics journal as . . . ," the most frequent student responses were "as a summary for me to study from later" and "as a record of the things I have learnt in maths."

Teacher Action.

- The most common student estimate of the frequency of teachers reading journals was "once a month."

- A Teacher Action measure was constructed from a cluster of related questionnaire items. The reported variation in Teacher Action with Teacher Identity was statistically significant, that is, the differences which students saw in the action which particular teachers took in relation to journal use were consistent and significant.

Mathematics and the School Mathematics Program

Since data were collected regarding student conceptions of the nature of mathematics and mathematical activity, the possibility exists for the later collection of parallel data in schools where mathematics journals are not in use. A comparative analysis of student responses may shed some light on the role of mathematics journals in developing particular student conceptions of the nature of mathematical activity.

- Students reported that their most common experience of mathematics at school was "listening to the teacher," closely followed by "writing numbers," "listening to other students," and "working with a friend."

- Students rated aspects of their mathematics course in order of importance. By far the most important was "the teacher's explanations." Other important aspects, in order, were: "the help my teacher gives me," "working with others," "my maths journal" and "the textbook."

The role of communication in the learning of mathematics and in the performance of mathematical activities was given considerable prominence by a significant majority of students. Pending further analyses, some sample student responses serve to illustrate the variety of student views about the value of mathematical journal writing. Responses include, "I find doing the journal

- useful, because it helps me explain to myself what I am doing wrong" (Year 8).

- hard, because sometimes you forget and other times you don't remember what you understood in class" (Year 10).

- a waste of time, because my teacher never collects my journal to help me" (Year 10).

- useful, because it helps me keep up with what's happening in class" (Year 12).

Teachers' Perceptions and Reported Practices

All eight teachers of mathematics responded to a questionnaire about their expectations of mathematics journal writing and the use that they made of journals. As a follow-up to the questionnaire, three teachers were chosen for interviews according to their experience in using journals and the year levels at which they taught.

There was a high degree of consistency among teachers' responses to the questionnaire. All teachers expected students to write in their journals for at least one hour each week, and to read over what they had written at regular intervals. Most teachers aimed to read all journals at least twice a term, with some expecting to do so more frequently, even though this was acknowledged to be a substantial time commitment. They also expected students to show journals to their parents.

For the majority of teachers, the most important thing for students to do in their journals was to write down what they understand. Likewise, a majority agreed that mathematics journals are most effective in showing how students think about mathematics. This was considered far more important than students' ability to summarize what they had learned.

A student in Year 7, for example, commenting on her review of place value and addition, said she was no longer learning how to do a "long sum," but learning "why I'm carrying." In her journal, she further noted:

> As many of us have worked on place value before, the object of this work is not to teach us how to do a long sum, but to do it so we understand. I must think about why I'm carrying one . . . is it a ten, a one (unit), or something else? . . . I must think about why I'm doing things with all sorts of maths, and not just do things automatically. That is how I was taught to do it.

Teachers tended to agree that students found journal writing difficult, and added that most students found it hard to explain what they thought. Journal writing was seen as helping students to write summaries, to be able to explain what they think, and, more importantly, to not be put off by mathematical words and symbols. One teacher commented that journal writing allowed students to investigate ideas independently.

When asked to consider the greatest benefit for students in reading over their journals, teachers very strongly believed that review of the journals was most valuable when students were trying to grasp a new idea. This outcome was rated more highly than using journals to go over material that has been dealt with before.

Teachers were more diverse in articulating the benefits they derived from reading students' journals. These ranged from getting feedback on teaching, identifying difficulties experienced by specific students, and seeing how students learn as well as seeing what students think they have learned. A very common response of the teachers was to view journals as a way for students to communicate to teachers their feelings about mathematics. In general, teachers consistently noted that reading journals had confirmed for them the importance of two-way communication as a part of mathematics learning.

Nearly all teachers saw themselves interacting regularly and often with students through their journals. These interactions most commonly took the form of writing comments in journals, talking to students about what they had written, and helping students to overcome difficulties they had mentioned in their writing, as well as suggesting ways in which students could improve the quality of their mathematical writing. A majority of teachers said that they often raised issues in class based on what they had read in individual journals. Several teachers said that they needed more time to read journals, to make comments, and to provide individual feedback.

When asked to be more specific about ways in which students could improve their journal writing, teachers consistently commented in favor of students writing more about their own thinking and asking more questions in their journals. These two responses had stronger support than "writing better summaries" or "collecting more examples."

When given the opportunity to say how they regarded the mathematics journals, the three universally endorsed responses were: as a way for students to communicate their mathematical thinking; as a record of students' difficulties in mathematics; and as a way for students to think through the mathematics they had done. All teachers agreed that reading students' journals had contributed significantly to what they knew about their students. Some specific responses were:

[The journals provide] a more precise indication of how much they understand.

[Through reading journals] I am able to identify students who have "no idea," or have difficulty expressing themselves.

[The journals] help students to clarify difficulties, and verbalize attitudes toward mathematics.

All teachers agreed that journals had helped them to understand their own teaching. Some specific responses were:

I used to dominate discussion. I now guide discussion and encourage my students to explore. . . .

They often say if I have explained something well or not. Easier to assess how well (or badly) you have covered a particular idea.

I now write notes on the blackboard in every lesson.

Despite their references to students' finding journal writing as challenging and, at times, a demanding task, all teachers affirmed that they saw improvement in students' journal writing during the year and, when viewed across several years, cumulative improvement. In the subsequent interviews, teachers were asked to explain what they looked for to indicate improvement in journal writing. The teachers' response to this and other similar requests was to offer illustrative examples from particular students' journals:

A student, described by her teacher as quite capable, at the start of Year 10, was using her journal to summarize, basically in her own words, what the teacher had written on the blackboard. Later in the first half of that year, she wrote: "I ran into a problem. When I do sums like this I need to . . . "

Towards the end of the year, having studied the effects of transformations on linear and quadratic functions, she began to investigate on her own the effects of the same transformations on sine functions: "I know what 'sine' looks like . . . . I'm surprised to find that the rules are similar to those for a quadratic function . . . . As I was unsure whether these rules apply to all or some functions, I went on to find evidence to support this claim."

In their questionnaire responses, teachers commented that practice in journal writing had enabled students to express ideas more clearly and to relate ideas; they also noted that students' summaries had become more detailed and accurate.

Another Year 10 student was described by her teacher as "just taking notes at the start of the year." At this stage, her journal was used mainly as a device to summarize work done in class. Later in the year, she began to make comments on her own work, such as, "I'm still getting confused on what numbers to use in the domain and co-domain."

Towards the end of the year, the same student wrote: "Before today, I didn't realize what f(x) meant. Today I learned that f(x) means function of x."

Her teacher annotated this entry, asking her to explain this comment, and suggesting that she should try to analyze her own thoughts further.

Finally, teachers were asked whether their view of mathematics journals had changed over the period they had been using them. Three of the eight felt that there had been no change, commenting that they had always supported the use of journals in mathematics. From other teachers, there was a developing sense of greater appreciation of the value of mathematics journals. Some typical responses were:

Journals are a more powerful tool than I once thought.

My appreciation of their benefits has increased, as has my ability to assist students to use them.

I am a lot more aware of their usefulness.

Teachers brought to the interviews several journals by students, representing a range of ability. From the interviews, it was clear that these teachers used consistent criteria to track improvement in the quality of students' journal writing. Improved journal writing was noted from individual students within a single year level and by a comparison among journals from students in the same class. The criteria used by teachers supported the classification of developmental stages in mathematics journal writing which has been employed in this study.

Progression in Journal Writing

Teacher interviews, together with an examination of students' journals, served to confirm the categories that had been used to classify the major developmental stages in students' mathematical journal writing. It appears that the three categories, Narrative, Summary, and Dialogue, as employed to categorize questionnaire responses, showed a marked consistency with the linguistic forms by which students communicated what they had learned and how they had gone about it.

In the Narrative stage, students' frames of reference for their journals are defined, in the main, by tasks which make up the mathematics lesson and by the chronology of the mathematics classroom. In some instances, the description may be as bald as, "Today we did the pink sheet." Students seem satisfied, at this stage, to describe themselves as, "doing fractions," or "in the middle of chapter 3," and to comment on their learning in general terms, such as "It was easy," or "I finished all the work and got most of it right." Examples seem chosen to do no more than illustrate the work done. Many students at Year 7, as they begin to use journals, may be expected to be at this Narrative stage. A teacher of middle secondary classes described many students at this level as still coming to terms with journal writing. They are either still at a Narrative stage or just beginning to move into a Summary stage. To use this teacher's own words:

It is a case of knowing that they have to write something, but many have difficulty knowing what to do. At the beginning of the year, these students are saying what they did in the mathematics class. . . . They are able to describe what they did, and the types of things they did.

Unlike students' writing in a narrative stage, there is now a deliberate effort to delineate key features of the territory. The mathematics may still be "out there," but students give greater attention to describing key steps in their work. It is no longer sufficient for students to describe in very general terms what they are doing; journal writing provides an opportunity for them to "map" the territory in some detail and to record their progress. However, their descriptions are almost devoid of personal commentary or reflection. At this stage, their frame of reference is restricted to recording, "in very basic terms," the mathematics that has been covered in class.

With further refinement, students begin to include themselves in their summary of the mathematics covered. Not only is there more detail about what has been covered in class, but they are beginning to locate themselves in relation to the mathematics being taught. They begin to identify "problems" in their own learning and to describe how they achieved a solution. It is common, at this stage, for students to illustrate their work by reference to several examples and by comments on them. Yet there is little discussion of why these problems arose and little analysis by the students of their own thinking.

At a more developed stage, students begin to focus on the ideas being presented. This term marks a significant transition in the frame of reference for students' journal writing. A threshold is crossed when students begin to relate the mathematics being taught to what they are learning and begin to demonstrate their ability to connect new ideas with what they already know. One does not simply record or summarize ideas. One has to try to make sense of, or, come to terms with them. They can be illustrated by, but are not identical with, examples. Ideas make up the territory, but no longer is the territory seen as fixed and unchanging. The student is part of the territory and can change the way it looks. Communicating ideas and connecting them to what is already known now become key features of students' writing.

At this stage, students are able to identify and analyze their difficulties, suggesting reasons why they are thinking in a certain way. According to teachers, students begin to question what they are doing and show increasing confidence in using their own words to link ideas. They are able to make suggestions about possible ways to solve problems, even if these approaches may not prove to be successful. They are able to talk more confidently about questions they have "in mind." Through their writing, they show that they are actively teaching themselves mathematics.

Teachers can play a critical role in helping students to assume this degree of control over their learning. Getting students to articulate their own thinking at the point where they are coming to terms with a new idea, or meeting difficulty, is essential to helping many to move into the more reflective mode of writing, characterized as Dialogue. The key is to encourage students to question themselves when they do not understand, rather than rely on the teacher to tell them whether they understand. As one teacher wrote in a student's journal: "Unless you can explain it to me, you don't really understand."

Articulating their own thinking in their own terms is challenging and empowering to students. As they move into this mode of journal writing, many students frequently comment on realizing "just how valuable the journal has become." Helping students to achieve this level of development in their journal writing was a goal which teachers saw as achievable by many students and to which all teachers expressed genuine commitment.

Students and Teachers: A Brief Comparison of Views

A comparison of student and teacher data is informative in at least three ways. Such comparison can reveal (a) the perceptions of the purpose and value of journal writing held by the two groups, (b) the extent to which teacher expectations are realized in student practice, and (c) the way in which student perceptions of teachers' actions and beliefs match the professed beliefs and actions of the teachers. From the emergent commonalities and differences of view, it was evident that, while the classroom implementation of journal use may not universally match the stated policy and goals of the school, both teachers and students saw real value in journal writing. The statements which follow summarize points of contrast and consistency between the two groups' accounts of journal writing:

- While most students reported that they were writing in their journals after every mathematics lesson, as required, the amount of time spent in this writing was typically less than teachers' expectations.

- Three-quarters of the student sample reported that their teachers read the journals at least as frequently as the "twice a term" which most teachers reported. Student data revealed that the frequency with which teachers read their students' journals was predominantly a characteristic of the individual teacher.

- Contrary to teachers' expectations, very few parents ever read their children's mathematics journals.

- Teachers recognized the difficulty many children experienced in trying to explain their thinking.

- Responses from teachers and students stressed the importance of communication as a part of mathematics learning.

- The regular interaction, which teachers saw as arising from journal use, varied substantially with student perceptions of the actions of individual teachers.

- Those aspects of journal writing which teachers most frequently saw as needing improvement focused on characteristics of the Dialogue mode. Student responses were more varied. The forms of improvement which received significant support from students were as diverse as their modes of use.

- Senior students reported an improvement in their journal writing. Teachers felt that students progressed in their writing in the course of a year. The proposed taxonomy of journal writing (Narrative, Summary, Dialogue) emerged as a robust, powerful, and informative model of this progression.

The use of student mathematics journals at Vaucluse College offered the possibility of communication in all three modes—communicating about mathematics, communicating mathematics, and using mathematics to communicate. In particular, the integrated development of communication skills and mathematical thinking was central to the aims of the Vaucluse program. For some teachers, the ultimate goal of journal writing was to equip students to use mathematical forms and structures to describe their everyday world. However, the nature of journal writing derived from classroom purposes, and this close connection with schoolwork may not have offered students the opportunity to extend their growing confidence in mathematical language by applying it to situations outside schoolwork.

SOME BROADER ISSUES

Communication in mathematics is not a simple and unambiguous activity. The significance of this study is that it points to modes of communication as indicative of stances towards learning mathematics and ultimately of students' perceptions of mathematical knowledge. The categories which we have employed serve a dual purpose: as descriptive of students' perceptions of their learning of mathematics and, in the second instance, as a progression in student mathematical activity.

When students write in the Narrative mode, they see mathematical knowledge as something to be described. In the Summary mode, students are engaged in integrating mathematical knowledge, now conceived of as a collection of discrete items of knowledge to be collected and connected. When writing in the Dialogue mode, students are involved in creating and shaping mathematical knowledge.

IMPLICATIONS AND DIRECTIONS FOR FURTHER RESEARCH

The IMPACT procedure is now in wide use nationally, having been applied in the teaching of students from Year 4 of primary school to third year tertiary, and several local—that is, school-based—evaluations of its effectiveness are being conducted.

The publication of *Assessment Alternatives in Mathematics* (Clarke, 1989) has received an enthusiastic response, and the implementation of its contents would significantly alter the quality and the diversity of the modes of communication typically employed in mathematics classrooms.

With regard to the use of mathematics journals, a critical consideration for other teachers of mathematics is the nature of the interaction and the communication opportunities which student journals offer. We hope to continue the Vaucluse Study and to report in greater detail on the teacher's role in nurturing the emerging dialogue and responding to signs of increasing student reflection and changes in the quality and sophistication of their communications. Comparison of the Vaucluse data with responses from students and teachers in other schools would shed further light on the possible effects of journal writing on student conceptions of mathematics, mathematical activity, and school mathematics practices and on the significance of communication in the learning of mathematics.

The student journals themselves constitute a unique data source on the way in which students construct mathematical meaning and on the developmental stages in students' ability to make such constructions. Our understanding of communication and the relationship between language and mathematics learning may also be informed by a more detailed study of the nature and process of journal writing.

11

Measuring Levels of Mathematical Understanding

Mark Wilson

When we think of the learner as an active participant in constructing his or her own conceptualization of mathematics, we are forced to reassess the nature of mathematics tests. Traditional tests were based on an atomistic model of knowledge. Newer tests, based on a model of developmental change in understanding, are needed. This article describes recent advances in developing such an approach.

One way to measure student achievement is to give a test and to record the questions answered correctly or incorrectly. In modern test theory (such as Item Response Theory [Hambleton & Swaminathan, 1985] or Rasch Model analyses [Wright & Stone, 1979]), a student's standing on an achievement variable is estimated from the resulting vector of right and wrong answers. This variable is calibrated and criterion-referenced by the test items that students attempt and so provides a framework for mapping student progress. If the aim of an instructional program is to provide students with an unstructured body of facts, skills, and algorithms, then this methodology can be particularly appropriate. Items can be constructed to indicate the presence or absence of specific pieces of mathematics on any given occasion, and students' performances on those items can be scored either right or wrong. However, not all curricula are based on the premise that learning is a matter of

213

absorbing and reproducing provided information. Another way to build a curriculum is to concentrate on the conditions under which students change the way they conceptualize a subject. Progress occurs when a student discards a less sophisticated model or representation of a phenomenon in favor of a more expert conception. Traditional mathematics achievement tests are not well suited to the identification of the conceptions that students bring to problems. A new testing approach is required to map progress in conceptual understanding. This article describes recent advances in developing such an approach.

UNDERSTANDING AS A CONSTRUCTIVE PROCESS

A view of learners as passive absorbers of facts, skills, and algorithms provided by the teacher is the basis of most current measurement theory and practice. Standard achievement tests measure students' abilities to recall and apply facts and routines presented during instruction. Some items require only the memorization of detail; other items, although supposedly designed to assess higher-level learning outcomes like "synthesis" and "evaluation," often require little more than the ability to recall a formula and to make appropriate substitutions to arrive at a correct answer. Test items of this type are consistent with a view of learning as a passive, receptive process through which new facts and skills are added to a learner's repertoire in much the same way as bricks might progressively be added to a wall. The process is additive and incremental: students with the highest levels of achievement in an area are those who have absorbed and can reproduce the greatest numbers of facts, formulae, and algorithmic productions. The practice of scoring answers to items of this type either "right" or "wrong" is consistent with the view that individual units of knowledge or skill are either present or absent in a learner at the time of testing. Under this approach, diagnosis is a simple matter of identifying unexpected holes or gaps in a student's store of knowledge. This creates a perceived need for remedial teaching that fills a deficit in those subareas of learning in which knowledge is "missing."

For some topics in the school curriculum, this approach to measurement may be appropriate. But recent research on student learning has led to a new view of the student as a constructive participant in building his or her own understanding of subject matter. Learners do not just absorb new information, but rather they construct their own interpretations and relate new information to their existing knowledge and understandings. Thus, experts and novices are seen to differ not merely in amount of their knowledge but also in the types of conceptions and understandings that they bring to a problem and in the strategies and approaches that they use. In cognitive science, comparisons of novices and experts in various fields of learning (Chi, Feltovich, & Glaser, 1981; Larkin, 1983; McCloskey, Caramazza, & Green, 1980) show that expertise typically involves much more than mastery of a body of facts: experts and novices usually have very different ways of viewing phenomena and of representing and approaching problems in a field. Expert-novice studies suggest that the performances of beginning learners often can be understood in terms of the inappropriate or inefficient models that these learners have constructed for themselves. Similar observations have been made in the field of mathematics education (Nesher, 1986; Resnick, 1982, 1984).

Expert-novice research does not in itself offer a panacea for the problems that arise from traditional views of learning, emphasizing as it does the differences between two (relatively) static states rather than the process of change, which should be the focus of assessment—but it does at least point out two end points of the process of learning. The importance of process in mathematics education has been emphasized in a number of surveys (D'Ambrosio, 1979; Freudenthal, 1983; Romberg, 1983), as have the active, constructive processes of conjecture (Schwartz, 1985) and problem solving (NCTM, 1980). A constructivist vision of what constitutes mathematics—the creation of (new) order—lies behind the epistemology of von Glasersfeld (1983) and Davis and Hersh (1981). The "conceptual field" approach of Vergnaud (1983) has also as one of its most important elements a constructivist perspective on how children's conceptions are built from problems they have solved and situations that they have met.

The "phenomenographers" in Sweden and other parts of Europe (Marton, 1981; Dahlgren, 1984; Saljo, 1984) have adopted a similar perspective, using clinical interviews to explore the different understandings that students have of key principles and phenomena in a number of fields of learning.

These interviews have revealed a range of student conceptions of each of the phenomena that these studies have explored and have illustrated the importance of forms of learning which produce "a qualitative change in a person's conception of a phenomenon" from a lower-level, less sophisticated conception to a more expert understanding of that phenomenon (Johansson, Marton, & Svensson, 1985, p. 235). Similar investigations on problem solving in both mathematical and science contexts has been carried out by Laurillard (1984). This interviewing technique has resulted in a conception of learning in which a student is considered to almost always have some understanding and some strategy when addressing a new problem. All learners are considered to be engaged in an active search for meaning, constructing, and using representations or models of subject matter. Rather than being "wrong," beginning learners have naive representations and frequently display partial understanding which they apply rationally and consistently. In arithmetic, for example, "it has been demonstrated repeatedly that novices who make mistakes do not make them at random, but rather operate in terms of meaning systems that they hold at a given time" (Nesher, 1986, p. 1117).

For the assessment and monitoring of student learning, an implication of this view of learning is that we must start measuring the understandings and models that individual students construct for themselves during the learning process. In many areas of learning, and in mathematics in particular, levels of achievement might be better defined and measured not in terms of the number of facts and procedures that a student can reproduce (i.e., test score as counts of correct items) but in terms of best estimates of his or her levels of understanding of key concepts and principles underlying a learning area.

CONSEQUENCES FOR MATHEMATICS ACHIEVEMENT TESTING

Traditional achievement tests begin with a statement of the instructional objectives to be assessed, which should be stated as directly observable student behaviors that can be reliably recorded as either present or absent (Bloom, Hastings, & Madaus, 1971). This advice tends to result in items that are discrete in their relationship to the objectives and involve relatively unambiguous performances. The epitome of this is the multiple-choice item, which, due also to its ease of use with machine-scored answer sheets, has made the multiple-choice item the automatic choice for test developers. Hence, the advantages of traditional achievement testing include (a) its provision for a close link between curriculum objectives that can be expressed in behavioral terms and the resulting measures of student achievement and (b) the specification of standard testing conditions and scoring rules, which reduce subjectivity in assessment and provide results that are comparable over time and across students.

However, a disadvantage of traditional achievement tests is that, because of the emphasis these tests place on precisely defined student behaviors, they can encourage students to focus their efforts on relatively superficial forms of learning. As Bloom himself wrote, such tests "might lead to fragmentation and atomization of educational purposes such that the parts and pieces finally placed into the classification might be very different from the more complete objective with which one started" (Bloom, 1956, pp. 5–6). Alternatively, one might base achievement testing not on the detailed specification of many observable student behaviors, each of which can be recorded as either present or absent, but on a consideration of the key concepts, principles, and phenomena that underlie a course of instruction and around which factual learning can be organized. This alternative approach recognizes that learners have a variety of understandings of phenomena and that some of these understandings are less complete than others.

The challenge, then, is to find out enough about student understanding in mathematics to design performances that will reflect these different understandings and to then design assessment techniques that can accurately reflect these different understandings. This is a much more theory-intensive test generation model than that used for traditional tests. Even in domains where much research has been done, it may be the case that important subgroups of students give responses that do not match our expectations well. Hence, the test development and implementation model that we need must allow for greater flexibility in item scoring and in interpretation of the test results.

The primary focus of a mathematics testing methodology based on an active, constructive view of learning is on revealing how individual students view and think about key concepts in a subject. Rather than comparing students' responses with a "correct" answer to a question so that each response can be scored right or wrong, the emphasis is on understanding the variety of responses that students make to a question and inferring from those responses students' levels of conceptual understanding.

One area of learning in which work has been done to understand how students think about and approach phenomena is the area of so called "open sentences." Take as an example the work of Sandberg and Barnard (1986). Elementary school students were asked to solve open sentences like $6 - * = 2$ and were then asked to explain their solutions. Sandburg and Barnard analyzed the protocols from these explanations to classify their solution strategies into one of the six types given in Table 11–1.

Table 11–1
Strategies for Solving Open Sentences

Strategy	Answer
1	Add all. When the form does not conform to the canonical structure add the two given numbers.
2	Interpret the operation sign as a direct instruction to perform the stated operation on the two givens.
3	Read and solve the problem from right to left when the equalizing sign is placed on the left.
4	Read and solve the problem from the right to the left when the problem first states the unknown.
5	Bridge the gap between the two given numbers. When the structure is not canonical then the difference between the largest and smallest number is determined.
6	Expert

The observations made in studies such as this one suggest that students do not simply make random "errors" but operate in terms of naive theories about mathematical phenomena. In the area of open sentences, Sandburg and Barnard (1986) found that "their answer pattern could be interpreted in terms of very systematic behavior. . . . Each child was found to use one overall strategy" (p. 5). Similarly, through their interviews with Swedish students about aspects of science learning, Johansson,

Marton, and Svensson (1985) arrive at a similar conclusion: "In our case, a discovery of decisive importance was that for each phenomenon, principle, or aspect of reality, the understanding of which we studied, there seemed to exist a limited number of qualitatively different conceptions of that phenomenon, principle, or aspect of reality" (pp. 235–36). Researchers have observed that the same naive conceptions can be found among students from different countries and with different educational backgrounds. Studies in four countries, for example, have shown that there is a systematic and understandable set of rules used by students who do not compare decimals in the standard way (Leonard & Sackur-Grisvald, 1981; Nesher & Peled, 1984; Swan, 1983).

Research findings such as these invite a reconsideration of the way in which we think about and attempt to measure student learning. Many students are succeeding on precise, operationally defined objectives without developing an understanding of the material that they are learning. Partial if not direct blame for this, at least for the ease with which this has become the norm, must surely be directed to the standards and practices that we have allowed to flourish in the testing community. For many mathematics educators, the answer is to place greater emphasis not on the learning of mathematical formulas and algorithms but on changing students' ways of thinking about mathematics. As one of the phenomenographers put it: "In our view, learning (or the kind of learning we are primarily interested in) is a qualitative change in a person's conception of a certain phenomenon or of a certain aspect of reality" (Johansson, Marton, & Svensson, 1985, p. 235). The assessment of such qualitative changes must equally become the goal of those who construct mathematics achievement tests.

LEVELS OF MATHEMATICAL UNDERSTANDING

A methodology for mapping student progress in conceptual understanding would first identify a variety of important concepts in an area of mathematics learning and then develop questions or tasks that can be used to explore the different understandings that students have of those concepts. A set of ordered categories would be defined corresponding to different levels of conceptual understanding within each task. The conception of ordered levels is basic to a view of learning as a "shift" or a "change" in a student's understanding. Such a change constitutes learning only if it involves a change from a lower-level, less sophisticated understanding to a higher-level, more sophisticated conception. Of course, there may be interesting conceptual changes that are not fundamentally ordered, and these need explication also, but such changes take on educational significance only in relation to students' progress toward more expert states (i.e., progression through the levels).

The set of (un-)ordered categories for a question is constructed by first exploring the variety of responses that students give when they are confronted with that question and asked to explain their thinking about it. To start with, the data from which ordered categories are constructed for a question are usually collected through student interviews. Qualitative analysis of the interview protocols results in ordered categories that provide a framework for recording future responses to that question and introduce the possibility of basing measures of achievement on students' levels of understanding. This is essentially the method used by Marton (1981) and his phenomenography group at the University of Gothenburg. There, researchers interview students to explore their understandings of particular concepts and principles, transcribe tape recordings of these interviews, and then carry out detailed analyses of transcripts. "The aim of the analysis is to yield descriptive categories representing qualitatively distinct conceptions of a phenomenon" (Dahlgren, 1984, p. 24). These categories form an "outcome space" that provides "a kind of analytic map" (p. 26) of students' understandings of each phenomenon. Learning is thought of as "a shift from one conception to another" (p. 31) on this map.

Returning to the example of open sentences depicted in Table 11–1, the strategy categories can be quite straightforwardly interpreted as ordered levels: level 0 is "no strategy"; level 1 is the use of strategies that are only sometimes successful, that is, strategies 1 through 5; and level 2 is the use of the expert strategy 6. In this interpretation, the structure of the levels would be identical for each item. Sandberg and Barnard point out that, in fact, the success of solution strategies 1 through 5 is dependent on which types of open sentence problem are being solved. For instance, strategy 1 will correctly solve an item like:$* - 2 = 7$, but strategy 5 will not. Under these circumstances it may be preferable to use a more complicated set of levels: let level 0 be "no strategy" as before; the incorrect strategies can be mapped onto level 1; the strategies that are correct in this case but not generally (i.e., not strategy 6) can be mapped onto level 2; and the expert strategy can be mapped onto level 3. An example of how this would work for three types of open sentence items is given in Table 11–2. This time the interpretation will be complicated by the fact that the strategies do not have consistent efficacies across problem types.

Table 11–2
Partial Credit Levels for Three Types of Open Sentence Items

Strategy	Item Type[a]		
	a	b	c
"No strategy"	0	0	0
1	2	2	1
2	1	2	2
3	2	1	2
4	1	1	2
5	1	1	2
6	3	3	3

[a]Exemplars of the item types are:
  a . . . $* - 2 = 7$
  b . . . $6 = * + 2$
  c . . . $7 - * = 4$

These interviews with students are essential for identifying the variety of understandings that learners have of phenomena and for constructing ordered categories for individual questions. But in many practical settings, interviews are not practicable as a basis for achievement testing. Alternative observation formats must be used for the purpose of assigning students to the categories that have been defined for test questions. This requires new kinds of imaginative tests that are capable of providing information about the conceptions that students bring to questions and that are also sensitive to the performance changes that can result from conceptual change.

One approach to exploring students' levels of understanding is through computer-administered tasks. When students enter their responses to questions into a computer, these can be matched to libraries of common responses that are keyed to strategy use. In this way, particular kinds of errors and misunderstandings can be identified and inferences made about students' levels of understanding. Clearly, in the open sentences example, it would be possible to assemble sets of open sentences that together would allow a decision to be made concerning which strategy was being used. Additionally, if a decision was not clear within a reasonable number of problems, this could indicate a student whose strategy use was either inconsistent, or of a nature different from the Sandburg-Barnard theory. Sufficient evidence of this type would lead to modifications in the theory itself.

A MODEL FOR MEASURING LEVELS OF UNDERSTANDING

The methods that have been developed for the analysis of right and wrong answers to test questions must be extended to support the construction of achievement measures from observations recorded in sets of ordered outcome categories. One such method, the Partial Credit Model (PCM), is described by Masters (1982) and Wright and Masters (1982); another, the Graded Response Model, has been described by Samejima (1969). Although these two models have certain important differences in terms of philosophical foundations and psychometric parametization, they yield quite similar results in practical applications. The PCM proposes that the probability of a person scoring in ordered level x rather than level $x - 1$ on a particular item i will increase steadily with ability in an area of learning such that the conditional probability of being in the higher category is:

$$\frac{\pi_{ix}}{\pi_{i(x-1)} + \pi_{ix}} = \frac{\exp(\beta - \delta_{ix})}{1 + \exp(\beta - \delta_{ix})}$$

where π_{ix} is the probability of a person responding in category x ($x = 1, 2, \ldots, m_i$) of item i, β is a person's level of ability in this area of learning (to be measured by this set of items), and δ_{ix} is a parameter that governs the probability of a response being made in category x rather than in category $x - 1$ of item i. By applying this simple logistic expression to the transition between each pair of adjacent outcome categories for each item, we form a connection between the ordered categories for that item and the underlying variable that the set of items is used to measure. This model can provide measures of achievement based on inferences of students' levels of understanding of each of a number of concepts or phenomena in an area of learning.

The PCM provides a framework for assessing the validity of attempting to summarize performances on the basis of different aspects of achievement in a single global measure. The PCM is used to construct a "map" to show how students' understandings of a phenomenon change with developing competence. In addition, the PCM provides a framework for identifying aspects of achievement in which a student is experiencing difficulty or making unexpectedly slow progress. The PCM takes as its basic observation the number of steps that a person has made beyond the lowest performance level. Consequently, the parameter to be estimated is the step difficulty (δ_{ix}) within each item. These step difficulties are substituted into the above model equation for the PCM to give a set of model probabilities for any given value of person ability. Figure 11–1 shows a plot of these model probabilities in a diagram called an "item response map."

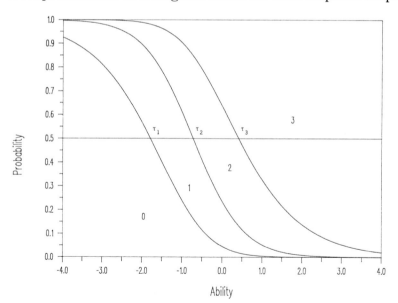

Figure 11–1.
Item response map (Item 1). (Kulm, 1990, p. 190). Reprinted with permission.

Responses to this item have been scored in four ordered categories labelled 0 to 3. In this picture, ability increases to the right on the page from −4.0 to +4.0 logits. The logit scale is a log odds scale. Thus, for a dichotomous item, the odds of success is calculated by taking the antilog (to base e) of the logit difference, and the probability of success is found by solving the equation L = log(P/(1−P)), where P is probability, L is the logit and log is the natural logarithm. For a person who is 1.0 logits above an item, the odds of that person succeeding on the item is exp(1) = 2.72, and the probability of success is exp(1)/ [1+exp(1)] = 0.73. This calculation is useful for gaining a "feel" for the interpretation of distance in the logit metric, but it must be emphasized that the interpretation for polytomous items is somewhat more complex. The best strategy is to read the probability directly from the Figure, as is done in the next paragraph.

From Figure 1 it can be seen that a person with an estimated ability of 0.0 logits (middle of the picture) has estimated model probability of about 0.05 of scoring 0 on this item; 0.18 of scoring 1; 0.42 of scoring 2; and 0.35 of scoring 3. The relative values of these model probabilities change with increasing ability so that, over the portion of the ability variable shown here, low scores of 0 and 1 become decreasingly likely, and a score of 2 on this item becomes increasingly likely up to an ability level of about 0 logits. As ability increases above this level, a score of 2 becomes less likely as the highest possible score of 3 on this item becomes an increasingly probable result.

The item response map in Figure 11–1 can be used to illustrate several important features of the PCM. Consider the horizontal line through the middle of the picture at probability P = 0.5. The intersection points of this straight line, labelled here τ_1, τ_2, and τ_3, are known in the psychometric literature as "thresholds." In dichotomously scored items there is only one threshold (or difficulty) for each item, defined as the position on the continuum at which the single ogive for that item intersects P = 0.5. One practical difficulty that arises in examining item response maps is that it is difficult to arrange more than two of them side-by-side in a reasonably sized figure. This is often required as the items are most often interpreted in relation to one another. The thresholds provide a way to summarize information about several partial credit items; simply place the

Thurstone thresholds next to one another on a "summary response map," an example of which is shown in Figure 11–2. A certain amount of detail is lost (in fact information is provided only about the points at which successive cumulative probabilities reach .5), but this is always the case with a summary and should not be a problem if the item response maps are provided as well. The Thurstone thresholds can be interpreted as the crest of a wave of predominance of successive dichotomous segments of the set of levels. For example, τ_1 is the estimated point at which levels 1, 2, and 3 become more likely than level 0, τ_2 is the estimated point at which levels 0 and 1 become more likely than levels 2, and 3, and τ_3 is the estimated point at which levels 0, 1, and 2 become more likely than level 3.

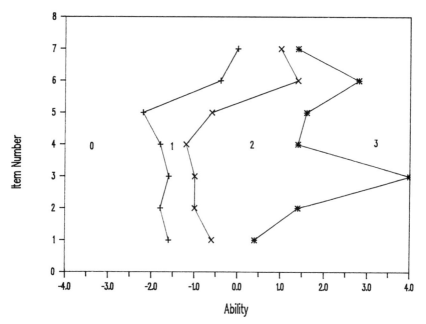

Figure 11–2.
A summary item response map. (Kulm, 1990, p. 192). Reprinted with permission.

The PCM makes no assumptions about the unconditional distributions of the persons along the latent trait but does assume that the model adequately fits the data. Tests of item fit identify individual items which function differently from other items and may lead to the conclusion that it is inappropriate to attempt to summarize all aspects of competence in a single measure. Persons who may be functioning differently from the majority are also indicated by fit indicators. Model fit can be assessed using a measure of fit called the "Item Fit t" for items and the "Person Fit t" for persons (Wright & Masters, 1982), which is a transformed mean square statistic. The distribution of these statistics is not precisely standard normal, so it will be used to focus attention on the more serious problems rather than to make a strict decision about whether persons or items fit or not. For items, it is possible to find the empirical item response map, which allows visual inspection of items that have been selected on the basis of the Item Fit t as questionable. Another way to assess fit is to divide the sample of persons into groups with interesting and interpretable differences, reestimate the parameters in each case, and examine the differences. Only if the model fits in the different groups can meaningful comparisons be made. These comparisons can be organized by using an indicator called the "standardized difference" between the estimates (Wright & Masters, 1982, p. 115).

EXAMPLE: USE OF THE SOLO TAXONOMY

The example discussed below is based on the Structure of the Observed Learning Outcome (SOLO) Taxonomy (Biggs & Collis, 1982), which is a method of classifying learner responses according to the structure of the response elements. The taxonomy consists of five levels of response structure:

1.   a prestructural response is one that consists only of irrelevant information
2.   a unistructural response is one that includes only one relevant piece of information from the stimulus
3.   a multistructural response is one that includes several relevant pieces of information from the stimulus
4.   a relational response is one that integrates all relevant pieces of information from the stimulus
5.   an extended abstract response is one that not only includes all relevant pieces of information, but ex-

tends the response to integrate relevant pieces of information not in the stimulus.

It is expected that in a given topic area learners will move through each level from the prestructural to the extended abstract as their comprehension and maturity improve. Furthermore, the majority of responses should be classifiable into one of the levels in the SOLO taxonomy indicating the learner's location on a latent dimension: "The structure of the SOLO taxonomy assumes a latent hierarchical and cumulative cognitive dimension" (Collis, 1982, p. 7).

In the particular items under study (Romberg, Collis, Donovan, Buchanan, & Romberg, 1982; Romberg, Jurdak, Collis, & Buchanan, 1982) a short piece of stimulus material, which might consist of text, tables, or figures, is supplied, and students are asked to answer open-ended questions concerning the material. Together, the stimulus material and the questions are referred to as a "superitem" (Cureton, 1965), and an example of one is given in Figure 11–3. The questions are linked to one of the higher four levels of the taxonomy. The responses are judged as acceptable or otherwise according to an agreed set of criteria, and the sum of the questions in a superitem is used as the indicator of SOLO level. In discussing the results, individual items within a superitem will be referred to as "questions" to help keep clear the distinction between levels. The following example uses data from a study of a new statistics curriculum for high schools (Webb, Day, & Romberg, 1988). In all, 1,238 responses without any missing data on the seven problem-solving items are available for the analysis. Because of the age of the students, only the first four levels (i.e., excluding extended abstract) are assessed.

In the case of SOLO superitems, the thresholds in Figures 11-1 and 11-2 can be interpreted in the following way. The first threshold, τ_1, is where it becomes more probable that a response will be unistuctural or above; τ_2 is where it becomes more probable that a response will be multistructural or above; and τ_3 is where it becomes more probable that a response will be relational rather than multistructural or below. In Figure 11-2, the unistructural threshold is marked by a "+", the multistructural threshold is marked by an X, and the relational threshold is marked by an "*".

1. The lines on the graph are city streets. One-way streets for vehicles are indicated by arrows.

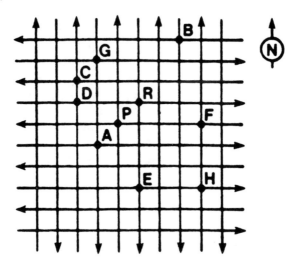

A. How many blocks would Alice (A) have to walk to visit her friend, Gayle, who lives at G, if she walks by the shortest way?

Answer _____

B. Alice (A) and Bill (B) have a friend Clara who lives at C. The three of them are walking from their homes to meet at a restaurant (R). Who has the furthest to walk?

Answer _____

C. If Bill (B) moves 2 blocks east and 5 blocks south, Gayle (G) moves 4 blocks south and 2 blocks west, and Alice (A) moves 6 blocks east and 2 blocks south, which person now has the farthest to go to the restaurant by car if the car takes the shortest possible route from each home?

Answer _____

Figure 11–3.
Item 1. (Kulm, 1990, p. 188). Reprinted with permission.

Results

The distribution of the students along the latent variable defined by the seven problem-solving items is shown in Figure 11–4, where ability has been estimated using the PCM (and is expressed in logit units). The great bulk of the students (90 percent) were estimated to be between −.63 logits and 2.38 logits (scores 9 to 18). Thus, in interpreting the item response maps, attention will be focused on this portion of the ability scale. Note also the nonlinear relationship between the logit scale and the scores; this is indicated by the selection of score locations given on the left-hand side of the figure. This will also need to be borne in mind when interpreting the item response maps. The analysis was performed using the *PC-CREDIT* program (Masters & Wilson, 1988).

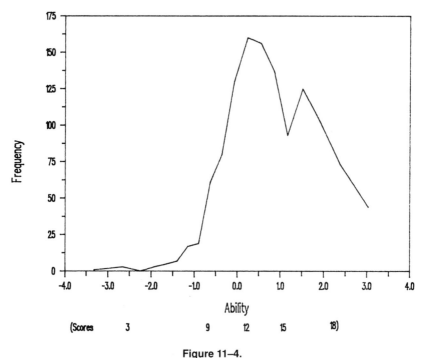

Figure 11–4.
Distribution of students along the problem-solving variable.
(Kulm, 1990, p. 188). Reprinted with permission.

Figures 11–2 and 11–4 give an overall picture of the progress toward a relational level of understanding that has been achieved by these students. All students are beyond the unistructural threshold for items 1 to 5, so, for these items, all are more likely to give a higher response than prestructural. Most students are displaying a level of understanding where prestructural and unistructural are less likely than multistructural and relational, but few give evidence of a mainly relational response. This summary response map leads to an interpretation of "typical performance" for students at a given location. For example, a student at 2.38 logits might be expected to respond thus to the seven items: (3,3,2,3,3,2 or 3*,3) (note that the estimated expected responses are real numbers, not integers—these "expected responses" have been rounded to the nearest whole number and are hence somewhat inaccurate). Overall, we would want to typify this student response as indicating a grasp of the relational level that is nevertheless challenged by the items with more difficult relational steps. Real student responses that correspond to a logit of 2.38 vary considerably from this expected pattern. For example, two student responses that actually were recorded were, for student A, (3,3,2,3,2,2,3) and, for student B, (1,3,3,3,3,2,3). Student A differs from the expected pattern only in one place, Student B, in two places. Are either of them seriously divergent from the expected pattern? The Person misfit indicator gives us some hint of this. It turns out that while Student A's pattern is quite innocuous as far as fit is concerned, the fit value for Student B is the highest recorded for any student. Thus, absent further information, one would be justified in making the above interpretation of the response vector for Student A, but one would need further information before one could make any such interpretation for Student B.

Item 1 was shown in Figure 11–3, and the estimated item response map for this item was shown in Figure 11–1. The conditional probabilities of response indicate that most students are performing above the prestructural level, ranging from approximately 15 percent of students with an item score of 0 (prestructural) at 9 points total (–.63 logits), to approximately 85 percent of students with an item score of 3 (relational) at 18 points total (2.38 logits). Thus, the predicted responses to this item range over the full SOLO spectrum within the range of ability of the majority of students in the sample. Moreover, the progress within the SOLO levels is quite regular from prestructural to relational for this item.

The relationship of the steps of Item 1 to the steps of the other items is also displayed in Figure 11–2, item 2, which concerns train timetable reading (Figure 11–5), has a pattern

*In this case, the expected value was approximately 2.5, hence 2 and 3 have about the same probability.

of thresholds similar to item 1 for the first two steps, but clearly has a more difficult transition to the relational level. The effect of this on the item response map for item 2 can be seen in the upper panel of Figure 11–6, where the band for score 2 (multistructural) is about twice as wide as that for item 1. Moreover, the relational threshold for item 1 is the easiest of all the items. It is interesting to consider the differences between the two "relational" tasks to try to understand this discrepancy. For item 1, the multistructural question requires the student to compare the distances from three separate places on a street grid to a fourth location; the relational question adds the complication that the three original places are moved. For item 2, the multistructural question requires the student to find the latest train that can reach a destination by a certain time; the relational question adds the complication that there is a certain time required at each end of the train journey for walking to and from the station. The item 2 question clearly demands that the student go beyond the immediate information provided by the timetable and use the timetable information in the context of a more complicated problem. The item 1 question uses different information to that provided by the original street grid, but the new information is of the same kind as the original— the student is asked to construct a revised grid. This is certainly more difficult than the multistructural question, but it does not clearly involve the understanding of a relationship among the pieces of information in the stimulus. What might a "taxi-cab geometry" item that was relational look like? Perhaps if the students were asked to use some standard geometrical concepts in the taxi-cab geometry world, such as, "What does a circle look like in taxi-cab geometry?", we might see more consistency between item 1 (Figure 11–3) and the rest.

Item 3 (Figure 11–7) displays a divergent pattern also, but this time the relational question is more difficult than that of the remainder of items. The lower panel of Figure 6 shows a very wide band for score 2 (multistructural), which makes a response on the relational level quite unlikely for this item. This item concerns the approximation of lengths of line segments to the nearest inch and half inch, using a ruler. The unistructural question requires the student to estimate the length of a line segment to the nearest inch; the multistructural question asks the same question, but specifies half inches; and the relational question makes this harder by misaligning the

A train leaves Alma and arrives in Balma at these times in the summer:

Leave Alma	Arrive Balma	Leave Alma	Arrive Balma
6:05 a.m.	6:50 a.m.	11:35	12:20 p.m.
6:55	7:40	2:08 p.m.	2:53
7:23	8:12	3:35	4:20
7:42	8:17	4:50	5:30
8:03	8:43	5:12	5:47
9:20	10:05	5:34	6:14
10:35	11:20	7:35	8:20

A. What is the latest train from Alma you can get if you want to reach Balma by 4:30 p.m.?

Answer _____

B. If you are busy working all morning and cannot travel before 10:00 a.m., what is the latest train you can get so as to reach Balma by 3:00 p.m.?

Answer _____

C. A person lives 30 minutes from Alma and has an appointment in Balma at 1:30 p.m. The appointment is 20 minutes from the Balma station. What is the latest time this person could leave home for this appointment?

Answer _____

Figure 11–5.
Item 2. (Romberg, Collis, Donovan, Buchanan, & Romberg, 1982).
Reprinted with permission.

line interval with the end of the ruler and failing to specify the standard (i.e., inch or half inch). Given this description, the distinction between the uni- and multistructural questions does not appear to fit so well into the SOLO framework. The relational question is obviously going to be harder for students, but this time it seems that the inconsistencies between this item and the others may be confusing students. This may be causing the relational question to appear very difficult.

Items 4 and 5 display a pattern similar to item 2. Item 4 concerns a survey of people attending a football game, and item 5 concerns the proportional mixing of liquids. As they, along with item 2, constitute the most generally consistent block of items, they will not be discussed at this point. Items 6 (item response map in top panel of Figure 11–8) and 7 (item response map in lower panel of Figure 11–8) exhibit a quite different pattern of thresholds from that of items 2, 4, and 5. For both patterns, the unistructural threshold is much more difficult

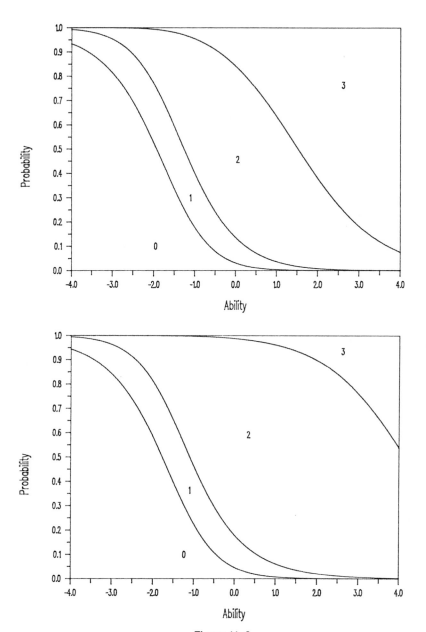

Figure 11–6.
Item response maps for items 2 (top) and 3 (bottom). (Kulm, 1990, p. 193).
Reprinted with permission.

3. When we use a ruler our measuring is not exact. To the nearest inch, the lines below are each 3 inches long. The lengths are somewhere in the range of 2½ inches to 3½ inches.

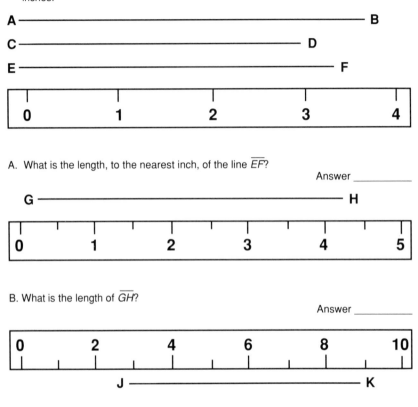

A. What is the length, to the nearest inch, of the line $\overline{EF}$?

Answer _____

B. What is the length of $\overline{GH}$?

Answer _____

C. What are the smallest and largest possible lengths of $\overline{JK}$?

Answer _____

Figure 11–7.
Item 3. (Romberg, Collis, Donovan, Buchanan, & Romberg, 1982).
Reprinted with permission.

than that of the other items, and the multistructural threshold is correspondingly harder also. This has resulted in item response maps that are "pushed" to the right compared with those for the other items.

Item 6 (Figure 11–9) has been criticized elsewhere (Romberg, Jurdak, Collis, & Buchanan, 1982; Wilson & Iventosch,

1988) on the basis of an ambiguous and relatively very complicated multistructural question, so this issue will not be pursued here. One would not expect this problem to make the
unistructural question unusually difficult—it is a seemingly
straightforward graph-reading question, although it does use
the word "average," which might mislead some students into
trying to calculate a mean. Item 7 (Figure 11–10) is a probability question about guessing the month and season in which
people's birthdates fall. Given some familiarity with probability,
the unistructural question seems a straightforward estimation
of an expected value. Perhaps the explanation of the discrepancy here lies not in possible misapplication of the SOLO taxonomy, but in the lack of familiarity of students in the sample
with statistics and probability. This would explain the transla-
tion of the Thurstone thresholds for the uni- and multistructural
questions towards the difficult end of the scale. The relational
questions in both cases do not experience so great a shift. This
might indicate that the lack of familiarity of the more able
students with statistics and probability was less marked than
that of the less able. This might be due to such topics being
customarily included in enrichment portions of curricula, or to
the possibility that students who are more able in general have
sufficient mathematical intuition and attention to detail to succeed on these items, where less able students need instructional exposure.

The fit of the items, as indexed by the Item Fit t, indicates
that the worst case, by a considerable degree, is that of item 1
(t = 5.78). The origin of this lack of fit can be examined by
considering the empirical item response map (solid lines in
Figure 11–11), constructed by calculating the proportions of
students at each total score that make up each item score and
then plotting them on an ability metric as was done for the
theoretical item response maps. Looking at this Figure alone
reveals two "blips" in the empirical map: one between –2.0 and
–1.0 logits, and a smaller one at about 1.0 logits. Some perspective on the meaning of "deviation" in this case can be
gained by superimposing the estimated item response map on
the empirical one. The dashed lines in Figure 11–11 show that
the theoretical response curves are very discrepant at the lower
end but tend to fit somewhat better at the top end, apart from
the second "blip". Notice how the theoretical curves tend to

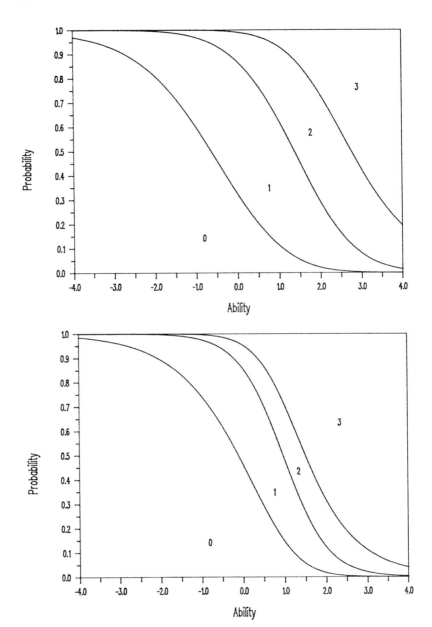

Figure 11–8.
Item response maps for items 6 (top) and 7 (bottom). (Kulm, 1990, p. 193).
Reprinted with permission.

6. The figure below shows the average birth rates, marriage rates, and divorce rates in Mapland for each 10-year period beginning in 1925 up to 1974.

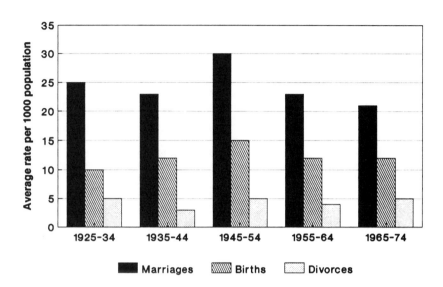

A. What was the average marriage rate in the years from 1925 to 1934?

Answer _____

B. Between which two periods did the average marriage rate decrease while the average birth rate increased?

Answer _____

C. What relationship seems to exist in general between birth rate and marriage rate?

Answer _____

Figure 11–9.
Item 6. (Romberg, Collis, Donovan, Buchanan, & Romberg, 1982).
Reprinted with permission.

7. A teacher tries to guess the season and month when any child in her class was born. If the teacher was to *guess the season*, she would most likely get 1 correct for every 4 guesses.
If the teacher was to *guess which month* any child was born, she would be likely to get 1 correct for every 12 guesses.

A. If the teacher used the *seasons* to make her guesses, how many times do you think she would have been correct with four children's birthdays?

Answer _____

B. The teacher has 12 girls and 16 boys in her class. She guessed the month in which each girl was born and the season in which each boy was born. In how many of her 28 guesses was she likely to have been correct?

Answer _____

C. If the teacher guessed 7 right out of 16 for the seasons and 6 right out of 12 for the months, how many more correct guesses altogether has she made than you would expect by chance?

Answer _____

Figure 11–10.
Item 7. (Romberg, Collis, Donovan, Buchanan, & Romberg, 1982).
Reprinted with permission.

balance between over- and underestimating the empirical curves above –1.0 logits. In calculating this statistic, greater weight is given to parts of the scale where greater information is available, so it is not necessarily the case that the greatest contributors to the statistic are the discrepancies that look greatest on Figure 11–11.

What has this told us about item 1? It looks uncomfortably like this item is susceptible to some sort of misinterpretation by students of lower abilities. The estimated step difficulties and thresholds are mainly being determined by the behavior of students of ability greater than about –1.0 logits. For students of lower abilities, their scores are being somewhat overestimated by these values. It looks as if some confusion occurs in students at about –1.0 logits that makes the questions relatively harder. Perhaps students who recognize the grid as being a Cartesian coordinate system (not the least able, obviously) make the problem harder for themselves by trying to solve for Euclidean distances. This is the sort of issue that can only be unravelled by gathering more information from students about their problem-solving tactics.

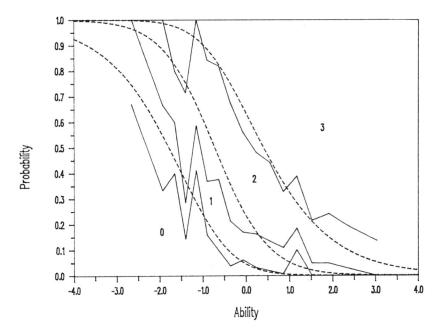

Figure 11–11.
Empirical and theoretical item response maps for item 1.
(Kulm, 1990, p. 194). Reprinted with permission.

As a comparison, Figure 11–12 shows the theoretical and empirical item response maps for item 3, which had a much better Item Fit (t = .50). For this item the considerable discrepancy at the lower end has not had so great an effect on the discrepancies at the upper end where most of the weight of the student information lies. This situation raises the possibility of an alternative interpretation of the large fit statistic for item 1. Perhaps the students at the lower end are simply less consistent about their problem solving than those over –1.0 logits, and the sum of these inconsistent responses for item 1 was one that, by chance, affected the estimation procedure. Unfortunately, there is no way to determine the most likely of these possibilities given the data. The empirical results and the analysis of them using an IRT model can show inconsistencies, but interpretation of such inconsistencies must be accomplished by probing more deeply into the students' cognitions than is revealed by scores on the items.

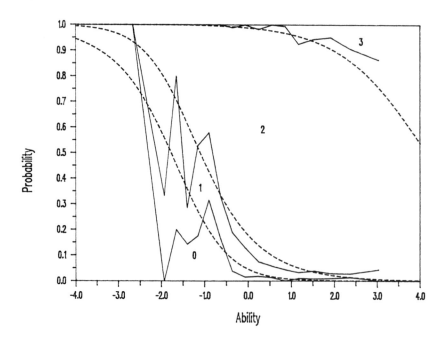

Figure 11–12.
Empirical and theoretical item response maps for item 7.
(Kulm, 1990, p. 195). Reprinted with permission.

Discussion of Example

The results for this example have included a "map" of the variable representing progress through the SOLO levels that allows one to give a criterion-referenced interpretation of a given student's mathematical understanding with respect to the SOLO levels and the items that were used to elicit performance. It also provided a framework for picking out performance patterns that were especially inconsistent and that deserve further examination.

The results have also pointed to some specific problems with particular items. Such results are not very useful if our attempt to measure levels of mathematical understanding is seen as a hit-or-miss, once-only task. If, however, measurement is seen as an incremental process, involving the gathering of information at various times in a variety of contexts, then the lessons learned from this analysis may be put to some good use. The empirical results can be used to sharpen the technique of translating the SOLO scheme into the reality of mathematical problem-solving items. The need to sharpen the distinction between multistructural and relational for one of the items was noted. Another item needs closer examination to clarify why its relational level is so difficult. One item displayed some inconsistency that may be due to confusion it caused for certain students. The best way to explore such empirical results is to collect samples of qualitative data at the same time as the item scores are recorded. This could be as straightforward as collecting a sample of the students' answer sheets (especially if they were encouraged to "show their work"). A more formal strategy would be to interview a sample of the students taking the test.

Looking more closely at some of the relational questions within the items (e.g., items 1, 2, 3 and 4) leads one to speculate whether the relational level has been well realized by these items. Certainly the relational question within each of these items would be expected to be more difficult, but that is not sufficient for it to be considered as indicating a higher level within the SOLO taxonomy. For example, in item 2, the relational question asks the student to place the use of a railway timetable into the broader context of a real-life problem where one has to consider time taken to get to and from the railway station. This is adding an extra variable to the problem, but is it addressing the mathematical relations among the components of the timetable? What is needed is a strongly mathematical idea of how to apply SOLO. One potential source for this is the van Hiele (1986) mathematics learning sequence. If one compares the SOLO idea, which is a general approach, to the van Hiele approach, one realizes that the van Hiele levels constitute successive relational levels that could be used in a SOLO framework. The interesting complication is that SOLO provides a framework for assessing within the van Hiele levels, and van Hiele levels provide a framework for linking between SOLO items at different levels.

12

A Framework for the California
Assessment Program to Report Students'
Achievement in Mathematics

E. Anne Zarinnia and Thomas A. Romberg

This chapter proposes categories for the California Assessment Program to use in reporting student achievement in
mathematics. Initially, state assessment reported achievement
to the legislature for the purpose of accountability. However,
assessment does more than simply register students' achievement; it affects it in ways both intended and unintended. In
these pages, the authors examine the explicit and tacit messages, imposed in the analyzing, gathering, and aggregating of
test data, that have subtle effects on teaching and student
achievement. It is determined that units of analysis and reporting categories are needed which will deliberately support
the purposes of adequate information for monitoring and—by
focusing attention on critical considerations—promote reform
in mathematics education.

With recognition of the impact of assessment and strong,
ongoing demand for educational reform, the goal of state assessment in mathematics is now to go beyond recording for
accountability purposes and to become an intentional catalyst
for educational change. Thus, units of analysis and reporting
categories are needed that will both deliberately support the
purposes of gathering adequate information for monitoring and,
by focusing attention on critical considerations, promote reform in mathematics education.

This document outlines seven bases for forming reporting categories. Each set of categories is associated with major issues, prevailing educational practices, and demands for reform. Also, each is discussed as it arises in the logic of the argument. The argument proceeds first by examining the present assessment situation in California as well as in other parts of the United States. In the course of that examination, four reporting categories are described. This is followed by a consideration of the primary objective of the reform movement—the development of "mathematical power" for *all* students. From that analysis, three additional reporting categories are presented, the last being recommended for use in California. Suggestions are made for gathering and reporting appropriate evidence. Finally, a recommendation is made based on a consideration of these alternatives.

THE PRESENT ASSESSMENT SITUATION IN CALIFORNIA

The gathering, reporting, use of, and reactions to assessment information, as these activities now occur in California, shed light on the problem of arriving at reform-oriented categories. There are a few key features to consider: first is the curriculum, intended, actual, measured, and achieved; second is the set of assessment and testing programs that are in place to measure and report the achieved curriculum; and third is the reaction of different groups of people to both.

For the last five years, the *Mathematics Framework for California Public Schools: Kindergarten Through Grade 12* (California State Department of Education, 1985) has been the state's outline of an intended curriculum. Each district has its own intended curriculum spelled out explicitly in its curriculum guide and tacitly in its textbook adoptions. In progressive districts, the guide has been revised to support the *Framework*. In other districts, the *Framework* may be mentioned but not really followed, or it may never be mentioned. The actual curriculum addressed by the teachers is undoubtedly a pragmatic mix determined under day-to-day circumstances.

Assessment is a means of reporting students' achieved curriculum. Whatever we know about the curriculum that students actually achieve depends on the way the assessment program measures and reports. The resulting information about the achieved curriculum is only good for accountability if what is measured is a valid proxy for the intended curriculum outlined in the *Framework*.

To grasp the dimensions of the reporting problem, it is critical to look at the broad picture of testing and assessment and at the ways in which program data from mandated testing is used. The National Center for Research in Mathematical Sciences Education has gathered information about the impact of mandated testing in the United States in a series of studies. The first surveyed the perceptions of eighth-grade mathematics teachers nationally (Romberg, Zarinnia, & Williams, 1989). The second surveyed state supervisors (Romberg, Zarinnia, & Williams, 1990). In the third (Romberg & Zarinnia, in press a), the issue was pursued through in-depth case studies in four states by interviewing teachers, testing directors, and administrators in selected districts. The fourth study (Romberg & Zarinnia, in press b) extended the pursuit of the issue to follow-up questions with teachers.

This series of studies has provided a base of information about the impact of mandated testing in California in two ways. First, a set of data for California was extracted from the national survey (Romberg, Zarinnia, & Williams, 1989). Second, California was one of the four states chosen to conduct the case studies and follow-up questioning for the third and fourth studies (Romberg & Zarinnia, in press a; in press b). California was selected because it has been actively pursuing educational reform by developing a state curriculum framework and by modifying its state assessment to support attainment of the standards in the framework.

The National Survey: California

Data (Romberg, Zarinnia, & Williams, 1989) suggested that California mathematics teachers are well informed about their state assessment program and perceive it as emphasizing mathematical understanding. In this respect, the perception of California teachers differs from the perception of teachers nationally that state assessment stresses essential competencies. The California teachers also distinguished quite clearly between the emphasis on understanding in the California Assessment Program and the basic skills nature of a typical district test.

The California Case Studies

The teachers in the national survey were selected randomly. That is not true of the districts in the case studies, which were chosen to reflect a range of curriculum and testing environments. Nevertheless, information from the case studies in several California districts can be used to develop a composite picture that illustrates ways in which existing state and district assessment programs exert an effect. The statements from teachers and administrators are used to write a coherent story and to point to issues that need to be considered in establishing reform-oriented reporting categories for mathematics.

Three assessment and testing programs have been mandated to measure the achievement of California's students. The federal government has required pre- and post-testing to select and account for *Chapter* 1 students. Second, the state requires that districts have a proficiency test for graduation. To satisfy this state proficiency requirement, school boards usually require students to take a standardized test. And third, California requires that every student must participate in the California Assessment Program (CAP).

District Testing. In one of the districts in the California case study, to ensure adequate performance on the Comprehensive Test of Basic Skills (CTBS) and achievement of minimum standards, all students are tested at least weekly in a Computer Managed Instruction (CMI) program and are required to master a series of objectives specifically correlated with the CTBS. Principals in the district are evaluated on students' performance on CMI and the CTBS. Their teachers refer to the mastery program as "computer-managed testing" and are skeptical about the *validity* of the CTBS in relation to their curricular objectives. To minimize the number of CMI tests to be taken by a student, the district administers the CTBS in the fall as well as spring, recording, thereby, mastery of as many of the CMI objectives as possible at the beginning of the year and reducing the time students spend in testing. The CTBS data are used to communicate with parents about their child's performance. They are also used to group and place students; one of the criteria for placement in high school algebra, for example, is a score of 80 percent or above on the CTBS. Although other districts are less intense in their approach, the uses of the district test are consistent with the data from the national survey.

The California Assessment Program (CAP). In addition to district testing with the CTBS/CMI system—or other tests in other districts—students are tested with the CAP in the spring using a matrix-sampling approach. Although each student takes only a subset of test items, within each school the total matrix addresses each of the subject strands and subtopics of the *Framework.* Results are reported by *Framework* strands. CAP returns a performance report to each school with school and state scores and change scores, as well as percentile ranking within a statistically similar comparison group of schools. In this way, schools can compare their performance with schools statewide, with schools in their statistical comparison group, and with their own performance in previous years.

Both teachers and school administrators in the case studies stated privately that CAP scores are important and are a basis for formal evaluation of administrators and informal evaluation of teachers. The pressure to use the scores in this way is considerable. CAP scores are printed in the newspapers in such a way that every school is compared with every other school.

The case study on the CTBS/CMI system revealed that the district administration requires principals to analyze and respond to the data in a written report that identifies perceived problems and outlines plans for dealing with them. Thus, the principals examine performance on the strands reported, pinpoint two to three weak categories or topics, and request the mathematics department to propose strategies for dealing with, for example, low scores in Measurement. An indication of the importance of the scores to districts is that although teachers in the district with the CMI system have no prep periods, they are released for a whole day to go to the district offices to discuss the CAP profile and review the mathematics program. These teachers regard CAP as helping them to get away from the overemphasis on the CTBS objectives, which they describe as almost entirely computational and to focus on the strands of the *Framework.* They said that CAP supports the *Framework* and validates their emphasis on problem solving. However, they also claimed a mismatch between what they do in problem solving and the efforts of CAP to assess it with multiple-choice items:

It has something that it calls problem solving, but I don't think the way that I do problem solving, you could put on a matrix type test, where you have just a few questions and it's multiple-choice. (Becker & Pence, in press)

The same teachers from the CTBS/CMI district were investigating alternative assessment strategies, such as portfolios, and proposed alternatives:

If we are going to have a test, I would like to see it be a collection of the different works the students have done, take a sample of their efforts to do a nice rich problem in the Fall, and then use that as a testing situation. . . . I don't know how they are going to go through life without being able to put all of these types of ideas together. (Becker & Pence, in press)

In another district, a teacher described a close match between the test and what he wants the remedial students to learn. He distinguished clearly between the kinds of problems he considered suitable for his remedial class and those suited to the accelerated class. He also had a different attitude about calculators for the remedial group, feeling that those students needed to know how to enter an operation into the calculator before they should be allowed to use one freely. Interestingly, the proficiency test—a dominant part of their mathematical experience for the remedial group, but a negligible concern for the accelerated students—does not allow calculators. The same teacher's accelerated students "behave as though their calculator is an extension of their hand." The irony is that many would consider this entrenchment of low-level approaches appropriate differentiation of the curriculum.

The pressures resulting from CAP's percentile rankings are a problem. A teacher in a district of lower socioeconomic status expressed frustration with the publication of rankings that minimize the impact of any improvement in scores. In a high scoring school, the percentile rankings are also bad for teacher morale and act as disincentives.

We are expected to be in the high 90s or they wonder why. Last year we scored 95 percent in math, and this year we scored 94 percent; we know our administrator is going to wonder why . . . and they'll be asking us to look over our program and see what area we are going to bolster up to get that score back. If we happen to come in the 93rd instead of the 96th, we did the best we could, the kids did the best they could. But at the same time, we get a feeling we are being criticized for it. . . . Well, if you got a hundred percent, you would be dead because the next thing you would be expected to do is improve beyond that . . . and you just can't. (Becker & Pence, in press)

In spite of this pressure, one teacher described the relationship between CAP, the 1985 *Framework*, and her teaching as follows:

Well, I have never seen the CAP test; I think here and there I have seen samples . . . to me it's just this big mysterious thing that I am really curious about. I am really trying to line up with that *Framework* because I think it's a very sound framework. I think it's very well balanced. I can't see too many things at this point that they would change. I think the trend now is to get more writing into the math curriculum, which still you can very easily slide that in. The *Framework* is making math fun to teach. I have taken lots of courses on using manipulatives, so I'm just very excited about it. . . . [CAP] is just as accurate, I guess, as anything else could be. But I still want more of an individual type score. I think that would be more helpful to us teachers. (Becker & Pence, in press)

Teachers interviewed for the case studies repeatedly said that the data from CAP are not very useful to them because of the time lag and because they do not get data for individual students. CAP is intended for program, not student, assessment. In fact, there are strong feelings in some quarters that CAP should not be extended to individual student assessment because of the embedded implications of state political control versus local authority. The fact remains that in lieu of measures of individual student achievement that represent a serious attempt to reflect the *Framework*, poor measures generated by the district tests are used:

My feeling is that the CTBS test is not updated. So, it's kind of out of sync with what is really happening. They are testing things we taught eight years ago. (Becker & Pence, in press)

Teachers in the CTBS/CMI district, which gives the CTBS twice and requires passing of the CMI objectives for graduation, commented:

The CTBS does affect what I do in the classroom, because I do have to spend time on those narrowly focused ideas. The test itself is almost defeating the ideas I have tried to instill in my classroom.

The CAP doesn't affect what I do; it does explain why I do it. (Becker & Pence, in press)

They observed that although the CTBS/CMI program is supposed to be diagnostic, it is not effective for that purpose, and they argued that their students "Flunk test-taking before they even have the opportunity to flunk content." One teacher complained that:

The kids who are poorest on the CMIs are the ones that can problem solve the best in class, especially if it is not a math-related problem. (Becker & Pence, in press)

Summary. In the case studies, both teachers and administrators subscribed to California's 1985 *Framework*. The districts met the obligations imposed by the state for proficiency testing, but teachers described the resulting emphases as computational. Teachers appreciate the efforts to correlate CAP with the *Framework* because it validates their efforts at problem solving. However, because districts did address the categories reported by CAP, it validated equally the problem solving conducted by teachers of accelerated classes and the computational emphasis of remedial teachers. Teachers decried the competitive pressures resulting from percentile rankings and were skeptical about CAP's ability to measure problem solving with multiple-choice approaches. They expressed a strong need for individual student data and some wanted alternative assessment measures.

Both districts and teachers need individual data. If CAP does not provide it, they will continue to use the most cost-effective data available. This is typically from low-level, district tests.

Conclusion

It is essential to recognize that information from testing and assessment programs is used for multiple purposes and that CAP is only one part of a coherent system. The need for data to support internal instructional decisions is paramount in the schools. Perhaps this is because instruction is the central mission of the system, whereas accountability is inherently an external issue. CAP does not provide individual student information, so schools derive that from other sources. The result is that although CAP is making strenuous efforts to adjust its program and move beyond basic skills and multiple-choice testing, it does so in the context of district tests that are substantially at odds with reform goals but that, nevertheless, are relied on for individual student data.

Every assessment, including those intended for program assessment, should provide timely information and consist of valid instructional tasks that are a conscious and integral part of the intended curriculum of each child. If this were in fact the case, there would be no need to distinguish between individual and program assessment with respect to appropriate tasks. Only the distinction between such things as sampling strategies and units of analysis would be significant. If one acknowledges student learning as the central mission of schooling, it further suggests that not only the tasks, but also the system and structures for gathering accountability information and reporting the data, should be designed with instructional needs in mind.

ALTERNATIVE REPORTING CATEGORIES WITHIN THE
EXISTING SYSTEM

Based on the description of testing in California and other current practices in the United States, four alternative approaches to reporting categories are apparent. These vary in the way items are categorized in terms of mathematical content and/or assumed abilities (or intellectual processes).

Alternative Number 1: Consolidate Testing and Update Content

Reporting Categories: Key Mathematical Content (Subcategories for process)

One potential response already under investigation in a number of school districts in California is a combination of CAP and updated district testing in an experimental program called Curriculum Alignment System/Comprehensive Assessment System (CAS). To replace low-level district testing in mathematics, a consortium of districts has attempted to alleviate:

- the massive amounts of time spent testing

- the lack of alignment between the curriculum and the tests

- the absence of individual student data in CAP.

A report from one of the districts is indicative of the process. The district formerly gave the California Achievement Test (CAT), as well as a criterion-referenced district test, and CAP; these three tests took about thirteen hours to administer. With the 1985 *Framework* in mind, the district now prioritizes its objectives, which are then mapped onto an item bank to generate the tests. The publisher sends a practice test, which the district administration says is an indication to the teachers of what they need to teach. On the actual test, the questions that compose the CAP matrix appear on the first few pages of a longer district test that is norm referenced. CAP items are returned to CAP, which generates the usual reports. The test publisher returns detailed analyses to the district for individual students and for classes.

The advantages of Alternative Number 1, in which the content of the tests is updated and merged for efficiency, is that it is nondisruptive. There are few major changes in strategies for gathering, analyzing, and reporting information. Through the district committee, the teachers can emphasize the aspects of curriculum that they value most and, thereby, ensure a closer match between the district test and the curriculum.

The disadvantage of this approach is that it is essentially an updating and consolidation of existing strategies. It is, therefore, unlikely to promote substantive change. It does not answer teachers' concerns about the use of multiple-choice items to test problem solving. Furthermore, CAP's traditional reporting categories do nothing either to change perceptions of mathematics or to focus on the measurement of mathematical power.

Alternative Number 2: Emphasize Mathematical Abilities in Each Content Area

Reporting Categories: Content (Reported only in subcategories of mathematical ability)

This second alternative expands the reporting of performance in each content area by sorting items into ability subcategories. For example, the National Assessment of Educational Progress has identified concepts, procedures, and problem solving as critical mathematical abilities (NAEP, 1988). Assessment for 1990 will report scores for abilities within each of the categories of content to be assessed. By reporting conceptual, procedural, and problem-solving scores and appropriately weighing items within each content category, NAEP hopes to emphasize the process of *doing* mathematics (see Figure 12–1). NAEP intends to use its assignment of content categories to reduce the emphasis on Number and Operations and to increase attention to Geometry, Algebra, and Functions. These tables illustrate the advantage of this alternative in the use of content categories to promote instruction in geometry (which has languished) and reliance on subcategories to clarify NAEP's vision of mathematics.

Table 1 Percentage Distribution of Questions by Grade and Mathematical Ability

Mathematical Ability	Grade 4	Grade 8	Grade 12
Conceptual Understanding	40	40	40
Procedural Knowledge	30	30	30
Problem Solving	30	30	30

Table 2 Percentage Distribution of Questions by Grade and Content Area

Content Area	Grade 4	Grade 8	Grade 12
Numbers and Operations	45	30	25
Measurement	20	15	15
Geometry	15	20	15
Data Analysis, Statistics, and Probability	10	15	15
Algebra and Functions	10	20	25

Figure 12–1.
Tables 1 and 2 from *Mathematics Objectives: 1990 Assessment.* (NAEP, 1988, p. 14).

However, the tables also illustrate the disadvantage of this alternative. It is likely that schools will use the categories for a multigrade summary analysis in which two or three weak points will be selected for attention. Despite the subcategories, the likelihood is that the primary emphasis of such analyses will continue to be on categories of content, doing little to change the vision in schools of what it means to engage in mathematics.

Alternative Number 3: Upgrade Process to the Same Status as Content Reporting

Categories: Content and Process

The problem of providing adequate information about performance is more deep seated than simply updating the content categories.

It has been popular to use content-by-behavior matrices. Such matrices have proven to be a powerful organizing structure. Despite modification of the specifics on each axis, the matrix approach has been used in many programs during the past quarter century. For example, it was integral to the model of mathematics achievement in the National Longitudinal Study of Mathematical Abilities (NLSMA) (Romberg & Wilson, 1969, pp. 29–44), and to all administrations of the National Assessment of Educational Progress. Persistence of the matrix as a tool for organizing activity is important and reflects:

- its power as an organizing tool;

- its visual facility; and

- the strong continuity between assessment projects created by relying on those with the most relevant experience in the field and those planning the next project.

Today, however, the inadequacies of this structure have begun to outweigh its advantages. Evidence for this lies in two basic areas. First, the content dimension remains unchanged. The result is implicit statements about curricula that focus on knowledge segmented into subjects for study, such as mathematics into arithmetic, algebra, and geometry. These have the immediate impact of implying that:

- knowledge can be broken down into clearly defined, independent, self-sustaining parts;

- such an approach is important, more important than any other approaches which might follow;

- there is a logical sequence of development in which each part builds on a preceding foundation;

- it is important to know about the divisions of knowledge enumerated;

- if knowledge were acquired in this manner, students would be able to use and apply their mathematical knowledge as needed.

Such implicit assumptions may be unwarranted if, for example, knowledge is regarded as unitary and emphasis is on knowing rather than knowing about. The approach is also unsuitable if there is genuine concern with application and problem solving. Stated simply, purpose should suggest form, and form implies purpose; incoherence may be inferred from anything less.

Disagreement over the precise structure and arrangement of content in a grid is only part of the problem. Westbury (1980) pinpointed a more fundamental concern: the difference between the intellectual structure of a discipline and its institutional structure in schools, where it is an administrative framework for tasks. The consequence is that administrative stability impedes intellectual change. For similar reasons, Romberg (1985) described mathematics in schools as a stereotyped, static discipline in which the pieces have become ends in themselves. A similar response to the impact of scientific management and behaviorism on mathematics as a school subject is Scheffler's (1975) denunciation of the traditional, mechanistic approach to basic skills and concepts:

> The oversimplified educational concept of a "subject" merges with the false public image of mathematics to form quite a misleading conception for the purposes of education: Since it is a subject, runs the myth, it must be homogeneous, and in what way homogeneous? Exact, mechanical, numerical, and precise—yielding for every question a decisive and unique answer in accordance

with an effective routine. It is no wonder that this conception isolates mathematics from other subjects, since what is here described is not so much a form of thinking as a substitute for thinking. The process of calculation or computation only involves the deployment of a set routine with no room for ingenuity or flair, no place for guesswork or surprise, no chance for discovery, no need for the human being, in fact. (p. 184)

A second concern is that the process dimension has been based on behaviorism. This reflects an application to the problems of education of the engineering approach to scientific management, focusing on managing environmental factors to achieve a defined outcome and ignoring the internal cognitive mechanisms. Scientific management rests on three basic principles: specialization of work through the simplification of individual tasks, predetermined rules for coordinating the tasks, and detailed monitoring of performance (Reich, 1983). These microprinciples pervaded American education with the same thoroughness with which they were applied in the economy. They dominated the breakdown of knowledge, the roles of teachers and students, instructional and administrative processes, the building-block approach of Carnegie units, the content and structure of textbooks, belief in the textbook as an effective tool for transmitting content, the structure of university education, and monitoring and evaluation. Hence, the notion of progress emerged through the mastery of simple steps, the development of learning hierarchies, explicit directions, daily lesson plans, frequent quizzes, and objective testing of the smallest steps, scope and sequence curricula.

Unfortunately, these are only the more obvious aspects. One consequence of such meticulous planning is that it renders the unplanned unlikely. A second is that a system designed to eliminate human error and the element of risk also eliminates innovation. A third is that, like factory work, it is dull, uninspiring, and unmemorable except for its boredom—for personal involvement and the mnemonics of the unexpected are nonexistent.

Bloom's *Taxonomy of Educational Objectives* (1956) epitomizes the domination of American education by scientific management, for it completed the process by which not only the content of learning but the proxies for its intelligent application were classified, organized in a linear sequence, and by definition, broken into a hierarchy of mutually exclusive cells. The consequences in the classroom were far reaching. Scope and sequence charts prescribed which parts of a subject were to be covered in what order; each cellular part of each subject was put into a matrix (e.g., Romberg & Kilpatrick, 1969, p. 285); behaviors suggesting desirable intellectual activity were also sequenced. However, given the multiplicity of subject cells to be covered, the easiest way to finish the prescribed course of study was to simply cover content without worrying too much about thought. Furthermore, matrices are difficult to construct effectively on paper in more than two dimensions. Consequently, few scope and sequence charts addressed both levels of thinking and specific aspects of content in a very coherent manner.

The dilemma such matrices pose for both assessment and instruction is whether to "cover" some content areas at all levels of behavior or to place emphasis on the lower levels of behavior for all content areas. This dilemma is partially recognized by CAP in its elevation of two process categories to the same vector as content: Problem Solving and Tables, Graphs, and Integrated Applications. Otherwise, CAP's *reporting categories* continue the content-by-behavior approach by reporting in major content strands and specifying sublevels for Skills and Applications.

CAP's strategy in mixing content and process categories in the same primary vector of the standard, two-dimensional framework recognizes that it is essential to focus on the process of doing mathematics. Therefore, processes to be valued as highly as content, such as problem solving, need to be elevated to the same category status in the reporting framework if they are to receive proportional attention in the curriculum.

The disadvantage is that elevating selected processes, such as problem solving, to the content vector effectively categorizes them as content. It implies that they are distinct from other categories of content, just as algebra is separated from geometry. Thus, although Alternative Number 3 supports California's 1985 *Framework* by emphasizing *problem solving, representation,* and *integrated application,* it only adds to the *content* of school mathematics and does little to support the significantly different view: that, in every area of content, *it is the doing of mathematics that is mathematics.*

Alternative Number 4: Focus Primarily on Process

Reporting Categories: Process (Subcategories for content)

If one wished assessment to convey the message that it is the *process* of doing mathematics—rather than simple coverage of content—that is closer to the nature of mathematics (e.g., Gale & Shapley, 1967) and the intent of the 1985 *Framework*, then the logical strategy would be to reserve the major reporting categories for mathematical processes. Thus, if one considered mathematics as problem solving, this would generate a set of major reporting categories originating in problem solving, such as the following:

Inquiring	Designing/ Modeling	Working Through/ Solving	Explaining/ Interpreting
probing	constructing	deducing	connecting
conjecturing	formulating	analyzing	communicating
		concluding	generalizing

CAP (1987a) adopted a similar strategy in the preliminary edition of its revised *Survey of Academic Skills: Grade 12,* which has two major reporting categories: *Problem Solving/Reasoning* and *Understanding and Applications.* The former is subdivided according to four major components of the problem-solving process: problem formulation; analysis and strategies; interpretation of solutions; and nonroutine applications/synthesis of routine applications. The latter is divided into the subcategories: number and operations; patterns, functions, and algebra; data organization and interpretation/probability; measurement and geometry; and logical reasoning.

Such categories have the advantage of representing mathematics as a purposeful and active occupation, especially if each of the processes is fleshed out in subcategories of process. They also address indirectly three of the five major goals of the NCTM (1989) *Standards:* reasoning, communicating, and problem solving. In addition, the notion of process categories is likely to seem very reasonable to those accustomed to the content-by-process framework because there is only a minor modification in thinking from the content-by-process approach. However, without content subcategories, there is little about process categories to distinguish them as unmistakably mathematical.

Thus, process categories have several disadvantages. First, if one were to treat processes as categories for separate courses or topics of study, it could lead to discrete textbook chapters and lessons on particular processes, such as *working through* or *solving*. There is no guarantee that processes would be assessed as part of a holistic task; in fact, a probable consequence would be to disintegrate problem solving in the same way that the meticulous definition of precise subcategories atomized content under the existing system.

A second disadvantage is that particular philosophies of mathematics generate different metaphors of mathematics as: problem solving; modeling; a cultural system; the science of patterns; a language. In fact, the power of mathematics lies in the fact that it is all of these (e.g., NCTM, 1989). Each metaphor contributes—and leads to—considerable insight into the nature of the mathematical endeavor. Mathematics as a science of patterns focuses on the discipline's search to identify and describe invariance and, consequently, on the big ideas of mathematics: quantity, space, dimension, chance, and change (MSEB, 1990).

Mathematics as a language emphasizes the discipline's universality, pithy symbolism, semantics and grammar, and generative nature. It also brings insight into the problems of those who are linguistically restricted or from minority cultures (e.g., Cocking & Mestre, 1988). If the analogy of study is pursued, tone, voice, clarity, and precision are all essential, implying the ability to represent one's beliefs about the beauty of fractals, for example, in a way and a medium that are appropriate to the argument and the audience. This philosophy leads to the belief that students should be able to convey their mathematical arguments in various representations, formally and informally, eloquently and appropriately (NCTM, 1989).

The crux of the problem is that if process categories are restricted to the integrity of a particular metaphor, they will probably fossilize and impoverish the vision of the mathematical endeavor in schools. If spread across multiple metaphors, they are likely to disintegrate the mathematical experiences of children as inevitably as minutely specified content has.

Summary

Each of the four alternatives proposed for reporting categories is viable, moderate, and attainable. However, each anticipates relatively minor changes in the existing system and portrays a restricted view of mathematics. If one takes seriously the task of measuring and reporting achievement in a way that encourages desirable change, the issue demands more than rearranging within extant structures. As they stand, the first four alternative structures for reporting categories are unlikely to bring about substantive change.

MATHEMATICAL POWER

The basic issues and goals of assessment need to be reconsidered and alternatives proposed that are powerful and practical and that make sense. The essential issue is that there is a considerable array of desirable information, but one can only have a limited number of reporting categories if their message is to be readily intelligible. This has led to the selection of a limited number of critical features for gathering and reporting information and resulted in a grossly simplified version of mathematics. The real problem is how to gather complex information and report simply and effectively, but not simple mindedly.

Epistemology and Authority

The single greatest issue in improving school mathematics is to change the epistemology of mathematics in schools, the sense by teachers and students of what the mathematical enterprise is all about. The magnitude of the current misunderstanding of mathematics is well illustrated by the fact that over 80 percent of the teachers who responded to NCRMSE's survey on mandated testing believe problem solving is included in their standardized district test of basic skills (Romberg, Zarinnia, & Williams, 1989).

The epistemology of school mathematics will be turned around only by its complete democratization and a change in the authority structure of the subject (Mellin-Olsen, 1987). The notion that mathematics is a set of rules and formalisms invented by experts that everyone else is to memorize and use to obtain unique, correct answers must be changed. In this context, there is an obvious potential for developing a unity involving the cultural genesis of mathematical ideas (Bishop, 1988), research in situated cognition (Brown, Collins, & Duguid, 1988), and a multicultural population of inadequately served minorities. The answers and the problems beg to be connected. Teachers, deskilled for decades, are pivotal. So, too, are students.

To lead school mathematics epistemologically, reporting categories must not only convey information about achievement, but carry clear messages about the nature of the mathematical enterprise, and every individual's role, rights, and responsibilities in the undertaking. The single most urgent message is political: mathematics is a universal, democratic, and collaborative endeavor in which all students are entitled to participate as citizens.

The question is, "What set of reporting categories might support this?" California's 1985 *Framework* introduced the idea of mathematical power as the goal of instruction:

> Mathematical power, which involves the ability to discern mathematical relationships, reason logically, and use mathematical techniques effectively, must be the central concern of mathematics education and must be the context in which skills are developed. (California State Department of Education, 1985, p. 1)

Despite introduction of the idea of mathematical power in the 1985 *Framework*, it is essential to recognize that the document focused heavily on outlining desirable mathematical content. Consequently, it is to its seven strands of content—Number, Measurement, Geometry, Patterns and Functions, Statistics, Probability, and Logic—that administrators in California make repeated reference. Their comments on the impact of testing suggest that they are also quite familiar with the *Framework*'s recommendations on instruction, especially with the emphasis on problem solving. This may be because problem solving is reported by CAP as one of the categories of content. In fact, when districts review their programs, it is the CAP categories that are addressed (Romberg & Zarinnia, in press a). One cannot assume that a focus on these categories will translate automatically into the mathematical power sought by the *Framework*.

The NCTM (1989) *Standards* also adopted *mathematical power* as the phrase most evocative of the quality of mathematical literacy sought for the entire population. Now, California is developing a new mathematics framework. Although still in its preliminary stages, it is clear the 1990 California "Mathematics Framework" will build both on increasing knowledge and conviction nationally about what it means to have mathematical power and on how students become mathematically powerful. The 1990 "Framework" will seek a new organization and structure of mathematics programs at all grade levels, one that emphasizes students' independent judgment and active engagement in mathematical investigation. It will also focus specific attention on assessment, both as an integral part of the classroom program and as externally imposed.

There are, therefore, five critical questions involved in the examination of mathematical power:

1. What is meant by mathematical power?
2. What can be regarded as convincing evidence of mathematical power?
3. How should that evidence be gathered and analyzed?
4. How should the evidence be summarized, and how can achievement of mathematical power be described in a report?
5. How corruptible is the resulting structure? If schools examine the report, focus on weak categories, encourage students to create evidence of improvement, and evaluate themselves by the instruments recommended, will students, in fact, become mathematically powerful?

Mathematical Power Interpreted

There is a strong distinction between definition of mathematical power as the intrinsic power of mathematics and the perception of mathematical power as individuals and societies empowered by mathematics. Therefore, one needs to think about what it means to be mathematically powerful both as individuals and as a society and to consider ways of identifying mathematical power.

All societies use and create mathematics. Mathematical power for the individual means that each person has the experience and understanding to participate constructively in society. Over the ages, people have invented and used mathematics to count, measure, locate, design, play, conjecture, and explain. They have also examined its generalized abstractions and developed out of them further mathematics—whether explanations, designs, proofs, or new theorems—which may or may not have been put to practical application (Bishop, 1988). They continue to do all of these, but in a rapidly increasing variety of contexts, in increasingly complex situations, and with shorter and shorter time spans for development. Assessment should seek and report evidence of these kinds of mathematical activity.

Mathematics is essential to value-guided optimization and choice between alternatives, for example. Consequently, understanding and experience of such uses of mathematics as precise and imprecise measures of vast and difficult to measure quantities is now critical to policy formulation and public decision making. In a society in which mathematics and information technology are pervasive, all members need to understand how mathematics is used; they need to know how to use it and to have a sense of how the discipline functions. The bottom line is whether we have a society whose members have a broad, reflective understanding and experience of mathematics in use, or whether we do not.

Mathematics is a profound and powerful part of human culture (MSEB, 1989, p. 33). It provides practical knowledge for everyday quantifying, locating, and designing. As such, it is the basis for science and technology and is deeply ingrained in aesthetics. Furthermore, in a culture that is heavily mathematical and technical, mathematical inference is at the root of rational argument and behind many debates on public policy. In the sense that citizens need a solid understanding of, for example, very large numbers, it is also a civic issue. Finally, it is a major part of the western intellectual tradition, and there is a deep vein of amateur mathematics in many leisure activities.

Everybody uses and relies on mathematics and, to some degree, everyone is a mathematician. However, not only do most students leave school with inadequate preparation, but "mathematics is the worst curricular villain in driving students to failure in school" (MSEB, 1989, p. 7). To counteract this situation, assessment programs should seek evidence of students using, reflecting on, and inventing mathematics in the context of value and policy judgments, designing and inventing, playing around with objects and ideas, describing and explaining their ideas and positions.

For a society to be mathematically powerful, its citizens must have the mathematical understanding and experience to jointly undertake the routine tasks of everyday life, operate as a society, and progress as a civilization. This means that in society both a critical mass of understanding and experience are needed and, in addition, availability of a substantial range of special expertise. Individuals have the potential for several kinds of power: to stand on their own with independent power; to contribute to the power of a group; to enhance and extend their own power by drawing on the group context. A recent trend by employers is the search for individuals who can work as effective members of a team. Hence, intertwined with individual power is the ability of a society to produce mathematically powerful groups. The full range of a society's power depends on the degree to which each of these facets exists in conjunction with the other. For accountability and encouragement, evidence of each facet should be reported.

The NCTM *Standards* argue that to be mathematically powerful in a mathematical and technical culture, students should develop the power to explore, conjecture, reason logically, and integrate a variety of mathematical methods effectively to solve problems. In becoming mathematical problem solvers, they need to value mathematics, to reason and communicate mathematically, and to become confident in their power to use mathematics coherently to make sense of problematic situations in the world around them. Hence, the document advocates four standards that should be used to critique all of the other standards: mathematics as problem solving, mathematics as reasoning, mathematics as communication, and mathematics as connections between topics and with other disciplines. Any assessment should provide evidence of each.

Students retain best the mathematics that they learn through construction and experience. Hence, the argument in the *Standards* is that students are more likely to become mathematically powerful if they learn mathematics in the context of prob-

lematic mathematical situations. As students use this approach
to mathematical content, they learn to formulate problems and
develop and apply strategies to their solution both within and
outside mathematics. In a range of contexts, they verify and
interpret results and generalize solutions to new problem situations. In so doing, they apply mathematical modeling and become confident in their ability to address real-world problem
situations. As they reason through their problem situations,
students develop the habit of making and evaluating conjectures, and of constructing, following, and judging valid arguments. In the process, they deduce and induce, apply spatial,
proportional, and graphic reasoning, construct proofs, and formulate counter-examples. Assessment of students' problem solving must reflect these considerations.

The problem situations referred to in both the *Standards*
and the draft of the 1990 "Framework" intend purposeful investigation of situations that are open to multiple approaches.
Students need experience in a range of prototypic situations so
that they can analyze their structure, finding essential features
and ways in which aspects are related. Prototypic is meant in
two ways: prototypic in the sense that the situation should be
representative of the kind of cultural context that has traditionally given rise to mathematics (Freudenthal, 1983) and prototypic
in the sense of the familiarity of the particular context to the
student. In the latter context, students need to be able to pose
a question, see the next question, evaluate a strategy, and
construct and discuss alternative methods. Having done so,
they need to examine assumptions and arguments and make
efficient choices.

To produce a worthwhile result, students may need to judge
what data are required, and then gather, process, and evaluate
them. They may also need to develop examples by which to test
conjectures. If the evaluation is unsatisfactory, they may need
to regroup for reconsideration. This suggests the need for fluency
with notational systems and the ability to develop abstractions
and explain clearly, to appreciate another's point of view, and
to arrive at a shared understanding.

Communication is essential to mathematically powerful individuals. In communicating with others about the problems
that they are engaged in, students develop the power to reflect
on, evaluate, and clarify their own thinking, to model situations, to formulate definitions, and to express ideas. In the

process, they discuss their conjectures with others and develop the power to make convincing arguments, to read, listen, and view with understanding, and to ask extending questions. Ultimately, their power to communicate will be judged by their versatility, fluency, and elegance in choosing, using, and switching between representations that both symbolize best the mathematical ideas under discussion and are most appropriate to their audience.

This is a picture of individuals who tackle problems with a confidence based on a combination of the coherent mathematical knowledge that has emerged from working experience and the collaborative support that comes from membership in a broader community. It suggests that both the power of mathematics and the mathematical power of individuals are multifaceted. All facets are required among members of a society, but different groups and group members may reasonably and productively be stronger in some facets than in others. The issue is how to gather evidence and report it in such a way as to set standards and describe range without imposing expectations on individuals that generate a sense of failure and lack of power.

This empowerment view of mathematical literacy differs from traditional conceptions in two major and inherently related ways. First, it goes beyond the typical stipulation of knowledge, skills, and application. *Power* carries connotations of control and authority as well as of driving force. To advocate that all students become mathematically powerful is to demand that they have the experience, confidence, desire, and independence to wield their knowledge actively and productively. It carries with it concepts of choice, judgment, initiative, self-evaluation, responsibility, collaboration, and mutual respect, almost all of which are missing in the present formulation of school mathematics.

Second, the change in language from purely symbolic problems to problems situated in a realistic context reflects the need for students to become immersed in significantly more complex, messy, and culturally based problems that are open to a variety of strategies and multiple solutions. The magnitude, or the unfamiliarity, of ensuing investigations may require extended effort by one student or the joint efforts of more than one student. The demand for problem situations is intrinsically related to the need for mathematical power. The demand for power, the recommended context of problem situations for developing expertise, extended projects, and the need for collaborative effort go hand in hand with the recognition that mathematics is value driven and value laden.

This kind of reflective, experience-based, collaborative knowing and doing is substantially different from the traditional pursuit of a sequence of independently acquired symbols, rules, procedures, methods, and skills whose sequential acquisition was presumed to aggregate coherently as effective mathematical knowledge. Standard algorithms may save time, but a student's standard algorithm does not have to be *the* standard algorithm. Note that coherence stems from purpose. A situation may cohere in several ways, depending on perspective. *It will have coherence for the student only if the perspective and purpose is the student's own.*

Definitions of mathematics as the science of patterns, as a language, as modeling, as a powerful abstraction, or as a tool for solving problems, will all continue to fall short unless students learn mathematics as something created by a community in which they are independent, collaborative, and contributing members. Their contribution may emerge either in response to a specific practical need, or tangentially from reflection, conjecture, argumentation, and validation. The root problem is to change the epistemology and politics of mathematics in schools.

Thus, the challenge is to cause—and gather evidence of—a radical rethinking in the classroom of what it means to learn mathematics. We are looking for students interested in thinking mathematically, purposefully, and productively rather than in accumulating an aggregate of classes that in combination purport to represent coverage of mathematics. The task of mathematics education is to enculture students into a democratic, entrepreneurial, mathematical, and technical society and to help them develop a sense of the culture of mathematics that is requisite. They must be empowered not only by a knowledge about mathematics but also by the confidence that, to some degree, they are mathematicians and members of the mathematics community. The immediate goal is to develop reporting categories to help communicate that challenge.

STANDARDS OF EVIDENCE IN THE IDENTIFICATION OF MATHEMATICAL POWER

If we expect students to use mathematics confidently and effectively to make sense of their world, we should gather and report evidence that they are using mathematics: *Direct evidence is essential.* If we expect them to use mathematics for design purposes, they should have design experiences and be assessed in that context. Similarly, if we expect them to play around with simulations or with abstract ideas in order to develop effective algorithms, hypotheses, explanations, or new mathematics, then it is in the context of these experiences that evidence should be gathered. This argument suggests that categories that would report the doing of mathematics, whether using or developing it, would elicit the most direct evidence of involvement in the mathematics process.

Alternative Number 5: Societal Uses of Mathematics
Reporting Categories: Societal Use

In reporting the condition of mathematics education in the United States, the Mathematical Sciences Education Board (1989, p. 2) outlined a series of societal uses of mathematics, offering a potential set of categories for the doing of mathematics.

Practical — knowledge that can be put to immediate use in improving basic living standards

Civic — knowledge to enhance understanding of public policy issues (A public afraid or unable to reason with figures is unable to distinguish between rational and reckless claims in public policy.)

Professional — knowledge as an occupational tool

Leisure — the knowledge and disposition to enjoy mathematical and logical challenges

Cultural — knowledge as a major part of our intellectual tradition

These categories have strong intuitive appeal for a number of reasons. First, they place immediate emphasis on mathematics use at all levels of society across major societal functions. They also make clear its pervasiveness. Each category is accessible and purposeful for all students. The set emerges from a study of the dimensions of the problem of improving mathematics education in the United States and leads readily to the use of problematic situations as the context for learning mathematics.

If, in conjunction with these reporting categories, an investigational and problem-solving approach were adopted, a student could engage in, for example, designing and furnishing of his or her bedroom within a given budget; analyzing and judging the arguments regarding the deforestation of Amazonia; drawing and analyzing the geometry and trigonometry of local Indian burial mounds; investigating the precise role of proportion in visual illusion; developing a paper on Mandelbrot, or speculating about numeric relationships. Such an investigational and problem-solving approach conveys a clear message about mathematics as something that everyone engages in purposefully and productively in the context of real situations.

These examples make a second feature explicit. The categories not only assume the integration of mathematics, they involve almost automatic connection with other content areas in the curriculum, with the student's personal life, and with the significant issues of everyday life. Selection of work in each of the categories can be tied readily to the individual and group interests of culturally diverse student populations. Each category is also amenable to efforts of different magnitude, whether group or individual.

The most obvious disadvantage is that, even more than with the process categories in Alternative Number 4, these categories have no overt relationship to mathematics and could be applied to any area of the curriculum. In addition, categories that focus exclusively on the uses of mathematics and ignore its invention implicitly leave the development to experts. This omission from reports on mathematical achievement for the entire population of students would impede democratization and fundamental change in epistemology.

Alternative Number 6: Cross-Cultural Genesis of Mathematical Ideas

Reporting Categories: Universal Human Activities That Have Prompted the Creation of Mathematical Ideas— Counting, Measuring, Designing, Locating, Playing, Explaining

A cross-cultural study of the genesis of mathematical ideas concluded that mathematics is a cultural technology that is invented by all societies. Every society develops the means to *locate* that is at the heart of geometrical and topographical reasoning. In each society, the most obviously mathematical activities are efforts to develop systems for counting and measuring. Design, both the abstract form and the abstracting process (the science of patterns), is at the heart of mathematical activity. It is essential to the development of counting, locating, and measuring systems, to the abstraction and invention of mathematical objects, and to the identification of more tangible objects and processes. Controlled, rule driven, speculative, and voluntary distancing from reality—briefly, play—is behind design and is essential to hypothesis development and modeling. Explaining focuses attention on the essence of the mathematical culture, the search for patterns that establish connections, and efforts to communicate their description effectively and elegantly. Communication is the representation of explanations. This distinction between explanation and communication becomes especially significant in the context of collaborative activity among a group of students (Bishop, 1988).

A set of categories based on this philosophy combines a focus on mathematics in use with emphasis on the invention and generation of mathematical ideas as a sense-making activity. The categories support reflective activity in the practical as well as the esoteric, fantastic, and theoretical sense. Furthermore, they lend themselves to the integration, rather than separation, of mathematical topics. This set of categories has the advantage of emphasizing mathematics in use and simultaneously being obviously mathematical. It focuses on mathematics as a discipline for making sense of the world and reasoning about itself without restricting the nature of mathematics to a particular philosophical metaphor.

In addition, the categories have the advantage over traditional designations—like abstractions, invention, proof, and application—of being more obviously based in cultural contexts while at the same time sounding both reasonably familiar and mathematical. The most important quality of this approach is that it emphasizes mathematics as a human activity undertaken by the great and small of all societies, individually and in cooperation, in accomplishments of tiny increments and huge leaps.

The set of categories selected for reporting should provide the information that the members of the educational system seek, whether parents, teachers, administrators, community members, or policy makers. It should support the traditional uses for accountability, placement, evaluation, guidance, and instruction. It should, above all else, support the two main concerns of providing information for monitoring and establishing a set of values for guiding change. In particular, it should answer such questions as:

1.   Do students think mathematically and use mathematics reflectively for practical and theoretical investigation in the course of authentic mathematical activity?

2.   Do students communicate their ideas with fluency, versatility, and appropriate technology in writing, speech, graphical representation, and appropriate symbols? Can they think on their feet, mathematically? Do they do these things adequately, effectively, accurately, and elegantly?

3.   Do students have a sense of mathematical community? Can they work with others? How do they function as part of an investigational team?

4.   Do they have a sense of the mathematical enterprise? Do they find mathematics valuable and fun? Do they have a strong sense of mathematical inquiry? Has the student's engagement in authentic mathematical activity engendered an enculturation in mathematics, a set of beliefs and understandings about the nature of mathematical knowing?

The need for direct evidence to answer these questions would suggest that the set of categories selected should address the doing of mathematics, for which Bishop's categories for the cross-cultural genesis of mathematical ideas are most appropriate. It also suggests that there should be categories to encourage collaborating, communicating, and developing a mathematical disposition. Thus, the following set of major categories is recommended to answer the educational system's most pertinent questions about mathematics education and to spur reform:

Alternative Number 7: Recommended Reporting Categories

Doing Mathematics: Locating, Counting, Measuring, Designing, Playing, Explaining

Representing and Communicating Mathematics: Mental and Represented Facility in Communicating—Verbally, Visually, Graphically, Symbolically

Mathematical Community: Individual Activity, Collaborative Activity

Mathematical Disposition: Valuing Math Confidence, Beliefs about the Mathematical Enterprise, Willingness to Engage and Persist

The first set of categories—Doing Mathematics—has subcategories that are not phrased in traditional mathematical terms. In other contexts, such mathematical terms as *space, number, logic* (Rucker, 1987), or *logic, number, measurement, space, statistics* (GAIM, 1988) have been used as a framework to describe mathematical activity. Unfortunately, the traditional terms that describe mathematical activity have become associated with a kind of school mathematics that is so sterile and divorced from the reality of mathematics—whether as a culture or in culture—that Bishop's (1988) alternatives are powerfully evocative. They are clearly close enough to traditional vocabulary for those who think in that language to make the connection, yet they emphasize mathematics in terms of active engagement, creative reflection, and productive effort. It takes little effort to see that *number* is included in the set of categories, but that *counting* emphasizes mathematical activity. *Play* is less obviously mathematical until one considers that intellectual "what-if-ing" in such domains as number, space, and logic is the essence of mathematical creativity. In fact, the subcategories of doing mathematics are essentially one classification of the mathematical problems that societies have addressed.

The categories are not, cannot be, and should not be, mutually exclusive. For example, explaining may be seen as communicating; in fact, one represents and communicates an explanation. Similarly, the categories for individual and collaborative activity are meaningless unless they apply to one or more categories of mathematical activity. Basically, the three categories for Mathematical Community, Communication, and Disposition would, of necessity, cut across all categories of mathematical Doing. In addition, one could not, for example, represent or communicate mathematics if there were no authentic mathematical activity to represent and communicate.

The intermeshing of the categories should contribute to an integrated and collaborative approach to mathematics education. First, with the exception of counting and measuring, the categories do not resemble existing expression—and, therefore, existing breakdown—of aspects of mathematics. Second, the categories are more open than traditional vocabulary to interdisciplinary effort and a more realistic mathematics education. Finally, when examined in conjunction with subcategories for individual and collaborative work, such distinctions as *explanation, communication, and representation* present clear opportunity for diverse individuals to make different kinds of contributions.

THE CONTEXT FOR GATHERING ASSESSMENT INFORMATION

However appropriate the categories for reporting evidence of mathematical power may be, they are corruptible. Changing one part of the assessment system does not guarantee that another facet of the system will not corrupt it. Suppose, for example, that the assessment of student performance in each reporting category were undertaken as a discrete act with discrete items designed especially and exclusively for a particular category—whether multiple-choice in format or not. The result would be to maintain the perception of mathematics as a collection of unrelated pieces. Similarly, if tasks are prescribed so that there is no student choice or initiative in the process and no opportunity for student self-evaluation, mathematics will continue to be seen as a compendium of expert knowledge to be covered. If mathematical discourse is assessed independently of authentic mathematical activity, it will be learned as a skill with slight chance for real application and transfer.

The Nature of Tasks

The potential corruptibility of any set of categories makes it clear that the attributes of the assessment task affect the validity of subsequent inference to such a degree that they must be considered an integral part of the evidence. This is already the case for very minor changes in wording between almost identical versions of multiple-choice questions and is also the basis for arguments of cultural bias in tasks. If one translates the traditional handshake problem from handshaking to kissing upon greeting, this is more evident: boys do not usually kiss boys; men in some societies kiss, but do it on each cheek. It is even more obvious when one is comparing students' performance on mental tasks, written tasks, discussion, or in alternate forms of representation. One cannot, with reliability, generalize from one task type to another.

In other words, task type and context are important qualifiers of evidence. They are the frameworks within which students can experience and demonstrate mathematical power. A rich task should be a microcosm of mathematical activity, open to student engagement in more than one—or either of several—categories of mathematical doing. Whether the task is generated by the student or the teacher, the quality and range of tasks and the nature of the mathematics addressed are strong indicators of the quality of the curriculum within which the student has experienced mathematics.

This is important because our efforts, to date, have been to administer a large number of small and relatively uniform tasks—which in no way could be regarded as microcosms of mathematical activity—in an attempt to describe performance in a specific cell of a matrix of competencies. This practice suggests that criterion-referenced tasks that are designed meticulously to elicit performance in one narrow domain provide evidence that can only be construed very indirectly as representing mathematical power. Time available for the task is part of the task context. Aggregation of performance on narrowly construed tasks undertaken under time constraints cannot be regarded as evidence of student power in more complex or time-elastic situations.

If you want to know what an individual can do in a difficult situation, under a short time frame, using a computer, either alone or as part of a team, you put the individual in a simulation of that situation and observe closely. If you want to know whether a student can talk, write about, graph, or present a logical argument in conversation about the mathematics of the task under consideration, you assess it directly by specifying context. The same is true regardless of whether the purpose is to find out if the student can produce a reasonable estimate under time pressure or if, after extended reflection, he or she can put forward a conjecture with an elegant explanation or use technology to create and investigate models.

An implication is that assessment tasks should vary with respect to time frame, familiarity, available technology, and social context. This inevitably means that some tasks might call for a simple and rapid decision, others for more contemplative judgment, yet others for extended and collaborative effort. In existing assessments, task type has been rigidly controlled and restricted. Measurement issues required narrow domain specification and economics dictated machine scoring—thus, the exclusion of open-ended answers and variable forms of representation. However, assessment of mathematical power requires a range of task categories and contexts in which the student is culturally comfortable. This is essential to the reporting and fostering of reform in school mathematics.

Few countries are so constrained in the tasks they set for students as ours is. Some of these following task types are being used in England, The Netherlands, and Australia (see de Lange, 1987; Department of Education and Science, 1985; Collis, Romberg, & Jurdak, 1986; School Mathematics Project, 1988):

- Extended Project Work (Individual and Group) (lasts about two weeks/five times per year) (School Mathematics Project [SMP], 1988)

- Open-Ended Tasks (Individual and Group) (GAIM, 1988) (last from 20 minutes to 90 minutes)

- Mental Facility Tasks (Last about 15 minutes; include judgments on spatial tasks, such as 3D rotations as well as rough or limited computation)

- Two-Stage Tasks (Initially undertaken as test items and then taken as homework for further exploration) (de Lange, 1987)

- SOLO Items (Multiple-test items linked to a single, more complex stem) (Collis, Romberg, & Jurdak, 1986)

As we reconsider the parameters of tasks in the context of efforts to change the epistemology of school mathematics, it is essential to acknowledge that the prescription of tasks implicit in outcome-based measurement (Sirotnik, 1984) detracts from that goal and also imposes culture. The School Mathematics Project recognized the importance of school-based, and student-initiated tasks—and, thus, local and individual freedom to control culture as well as other aspects of context—in its criteria for designing and evaluating extended project work (SMP, 1988). It opted deliberately for templates to analyze extended tasks which recognize the right of students to not only choose, but to conceive the focus of their own tasks. A similar move in California would restore local control while maintaining national standards and at the same time contribute to the democratization of the curriculum.

The Evaluation of Work

If one really subscribes to the idea that a change in the authority structure of school mathematics is essential to real change in its epistemology, that authority must be seen to transfer from external experts to the school, the teacher, and the student. The assessment process, including its tasks, is a key part of that process. As long as assessment is entirely an external dictate rather than a collaborative effort, the final reality for students is that they must learn somebody else's mathematics as opposed to holding their own mathematical ideas up for cooperative assessment by the total mathematical community, which includes their peers.

One effective strategy for democratizing school mathematics would be for tasks to include a strong element of choice, self-evaluation, and peer review. Student self-evaluation and peer review could be moderated either by the teacher, or on a sampling basis, by the assessment program. Self-evaluation and choice would serve the dual purpose of changing the authority structure and fostering the habit of self-analysis.

There is an additional advantage: choice allows identification of extraordinary achievement, whether in depth or range. Without choice, there is a serious problem of identifying appropriate range and level of specificity in the reporting categories. With choice, the range of mathematics undertaken by students can flex, and thus can be described more precisely.

The Strategy for Collecting Evidence

Each assessment task should be a sound instructional activity. Given that direct evidence is essential, alternatives to existing assessment tasks are required: There are several reasons why this should be the case:

a. It is quite obvious that the alternative assessment tasks described or envisioned (Stenmark, 1989) take considerably longer than the typical timed test. If an adequate number and range of tasks is to be engaged in, it makes sense for them to be an integral part of the curriculum.
b. One of the expressed intents is for the assessment to affect teaching. Hence, developing tasks that look as much as possible like the kind of teaching they intend to encourage would be the fastest way of producing an impact.
c. If tasks are sound instructional activities and take considerably longer than standardized test items, they should be incorporated into instruction. This implies continuous assessment, which serves to have a greater impact and simultaneously provides teachers with more usable information about students than is derived from standardized tests. It also provides more authoritative information for regular communication with parents.
d. If there is a major element of choice, self-evaluation, and school-based administration and analysis, the strategy will effectively and rapidly reskill the teachers, one of the biggest single challenges of reform.

There Should Be Agreed Aspects

Categories of tasks, common categories for analyzing evidence, common standards for judging evidence, a standard language of description, and formal organizational strategies for arriving at interjudge agreement need to be decided upon. Direct evidence is essential, the context is an essential part of the evidence, and common sense suggests that the assessment tasks be incorporated into regular instruction. However, additional problems need to be resolved if the reporting categories are to be effective. The first problem is that of finding an alternative to standardized tests for the purpose of accountability, a matter of valid assessment. The second is one of describing the quality of performance.

It is common knowledge that the high school of origin—as opposed to grades or test scores—is the most significant predictor of students' success in college. Standardized testing proliferated because transcripts from a particular school could be relied on to reflect little more than credit accrual and seat time (Wiggins, 1989). Thus, regardless of whether assessment tasks are a one-time affair or become part of instruction, it is essential that there be universally respected interjudge agreement if categories that rely on school-based assessment of such things as student portfolios of extended project work are to be usable for monitoring and reporting students' mathematical achievement.

Interjudge agreement, whether on a large or small scale, requires common rubrics for defining, undertaking, and analyzing tasks. The SMP (1988) assessment sheets for open-ended tasks (Figures 12–2 and 12–3) illustrate the use of rubrics to guide analysis, set standards, and enable the process of interjudge agreement. The investigational sheet in Figure 12–2 guides teachers through the process of assessing a student's mathematical investigation of a pool table. It makes clear what the agreed-upon criteria are for arriving at a grade, it identifies the teachers' task-related modification of the criteria, and it measures the student's assessed performance. Figure 12–3 demonstrates a similar set of rubrics and place for teacher comments for a student who conducted a practical project investigating the design of packaging for Smarties (similar to M&M candies). In addition, during the course of the project, the teacher can make notations and comments on the student's diary of the project. To arrive at reliable coding, teachers receive advanced training and also meet collegially with other teachers in their school and region to discuss grading. The strategy both enables teachers to arrive at interjudge agreement and also communicates the kind of mathematics sought.

The SMP (1988) strategy for logging, commenting, judging, coding, summarizing, and arriving at agreement about student work makes possible the incorporation of assessment as a routine and productive part of instruction. It also provides a basis for the development of a common language for describing the

Figure 12–2.

SMP Assessment Sheet for Investigational Project (SMP, 1988, p. 5).
Note: This grid is backed up by substantial written material in an accompanying handbook.

SMP 11–16
ASSESSMENT SHEET FOR OET
(INVESTIGATIONAL)

Candidate's name: ____

Brief title of task: ____

	Grade
	Final mark
	Teacher code

Stage	Process	Code	The pupil at grade F will be able to	The pupil at grade C will, in addition be able to	The pupil at grade A will, in addition, be able to	Teacher's comments in addition to shading boxes
IDENTIFYING	QUESTIONING/ EXTENDING (linked with EX)	QE	Decide the relationship to be established, the features to be investigated, and information to be obtained.			
	PLANNING	PL	Decide on the steps to be carried out and the order in which to carry them out.	Adopt a methodical approach by selecting variables and deciding which are important or can be ignored.	Apply reasoning to plan the approach and determine what might go wrong or recognise key results and use them to structure subsequent work and extensions.	
IMPLEMENTING	GETTING STARTED/ SIMPLIFYING	GS	Explore by looking at some particular cases.	Structure a simple start in order to approach a difficult task.	Use efficient methods to simplify a complex task.	
	WORKING SYSTEMATIC-ALLY	WS	Find and list all possibilities in a simple situation by haphazard trial and error.	Develop a system to find all the pertinent data in some of the cases used.	Work with a system which leads directly to generalising or proving.	
	CLASSIFYING	CL	Categorise information according to given criteria.	Decide to categorise information in relation to chosen criteria.	Produce general classification.	
	SYMBOLISING/ RECORDING	SR	Record results in a simple diagrammatic or tabular form.	Decide to record using conventional symbolic representations or novel diagrams.	Make use of a novel symbol system in an elegant manner.	
	CONJECTURING/ GENERALISING	CG	Recognise, describe and extend patterns.	Make a conjecture about a relationship and attempt to verify.	Make and test conjectures and formulate general rules.	
	CHECKING/ PROVING	CP	Check that a pattern applies to all data available.	Predict further case in order to check a generalisation or prove a situation is not possible.	Prove a generalisation by analytic explanation.	
REVIEWING	SUMMARISING	SU	Summarise the results and describe some valid observations or interesting features	Draw some valid conclusions.	Interpret the results achieved concisely including the key valid conclusions.	
	COMMUNIC-ATING	CO	Explain orally and in writing the problem, the route to the solution and the outcome using a step by step approach.	Give a clear general description, orally and in writing, of the progress with, and outcome of the task.	Give a clear verbal and concise written account of the task including the assumptions made and the strategies used.	
	EXTENDING (linked with QE)	EX		Describe or follow further enquiries to extend the scope of the work.	Explain any limitations to the work and relevant further enquiries, modifications or extensions.	

1992 ENTRY

© SMP

Figure 12–3.
SMP Assessment Sheet for Practical Project (SMP, 1988, p. 6).

Note: This grid is backed up by substantial written material in an accompanying handbook.

SMP 11–16
ASSESSMENT SHEET FOR OET (PRACTICAL)

Candidate's name	Brief title of task	Grade / Final mark / Teacher code	Teacher's comments in addition to shading boxes

Stage	Process	Code	The pupil at grade F will be able to	The pupil at grade C will, in addition be able to	The pupil at grade A will, in addition, be able to
IDENTIFYING	ANALYSING	AN	Choose some features which form sub-questions for enquiry.	Structure some prior analysis of the problem in generating sub-questions.	
IDENTIFYING	PLANNING	PL	Decide on the steps to be carried out and the order in which to carry them out.	Adopt a methodical approach by selecting variables and decide which are important or which can be ignored.	Apply reasoning to plan the approach. Determine what might go wrong, or recognise key results and use them to structure subsequent work.
IDENTIFYING	MODELLING	MO	Make some simple assumptions so as to be able to proceed.	Make considered assumptions and justify them.	Formulate a simple mathematical model.
IMPLEMENTING	EXPERIMENTING/ QUESTIONING	EQ	Design simple experiments and/or questionnaires.	Design systematic experiments and/or questionnaires.	Structure questions so as to produce data which can be immediately processed and leads directly to valid decision making.
IMPLEMENTING	SAMPLING/ COLLECTING	SC	Collect data from relevant sources.	Collate, tabulate and compare data from a variety of sources.	Use appropriate sampling techniques to collect data.
IMPLEMENTING	MEASURING	ME	Use measuring instruments relevant to the task.	Use measuring instruments to the appropriate degree of accuracy.	Consider limitations caused by level of accuracy of measuring instruments.
IMPLEMENTING	PROCESSING DATA	PD	Process data, carrying out simple calculations to produce a conclusion.	Process data after first ensuring that it is consistent in form e.g. comparing like with like.	Process data discriminating between necessary and redundant information.
IMPLEMENTING	REPRESENTING	RE	Produce simple sketches, graphs, diagrams, plans or scale drawings as appropriate.	Use a range of plans or scale drawings including, where appropriate, 2-D representations of 3-D situations.	Invent novel representations of complex 3-dimensional situations.
REVIEWING	CHECKING/ OPTIMISING	CH	Check reasonableness of answers.	Attempt to verify and relate answers to original conditions.	Obtain optimum answers in relation to original conditions.
REVIEWING	FORMULATING A SOLUTION	FS	Form some overall solution to the whole problem.	Formulate a structured solution involving relevant answers to sub-questions.	Produce an optimum overall solution to the whole problem discarding the trivial or over-complicated answers to sub-questions.
REVIEWING	COMMUNICA-TING	CO	Explain orally and in writing the problem, the route to the solution and the outcome using a step by step approach.	Give a clear, general description orally and in writing of the progress with, and outcome of the problem.	Give a clear verbal and concise written account of the problem including the assumptions made and the strategies used.
REVIEWING	INTERPRETING	IN	Comment orally or in writing on the limitations of the work done.	Interpret the outcome of the work done in the context of the original problem, describing limitations that arise.	Explain limitations in work done suggesting modifications or relevant further enquiries.
PROCESSING	ADAPTING	AD	Follow a rudimentary plan.	Follow original detailed plan.	Adapt starting plan flexibly as work continues.

© SMP

1992 ENTRY

tasks and the quality of achievement, which is essential if instructional assessment is to serve accountability purposes. Thus, for example, conclusions about the quality of verbal communication would result from the use of rubrics ranging from sloppy to precise and rigorous, and from inarticulate to articulate and elegant to analyze the quality of mathematical communication about a design project. Definition for, and by, teachers of the quality of communication that would satisfy each aspect would do much to set and implement standards among the participating schools.

In summary, rubrics for analysis and scoring of alternative assessment tasks would help ensure uniform judgment of quality, raise standards, and promote interjudge reliability. Furthermore, teachers cooperating to create and use agreed rubrics for commenting on student work would, in essence, integrate assessment with instruction. In a context in which they were reasonably sure of the reliability of their judgments against those of other teachers, they would also have their own grading for valid and timely information for instructional decision making. At the same time, the formal rubrics and strategies for accomplishing interjudge agreement would support development of alternative assessment tasks, prompt efforts to improve the mathematics curriculum, and help instill public confidence in the use of school-based information for accountability.

Multiple measures (NCTM, 1989) and an agreed-upon language permit significantly more informative statements about mathematical power, such as:

- The student invents elegant algorithms to design three-dimensional movement of robot arms, often in familiar contexts.

- In group endeavors, the student suggests and engages in significant algebraic generalization and conceives extensions, usually in the familiar context of dairy farming.

- Sixty percent of eighth-graders in Lodestone, California, clarify assumptions about their projects; a few of them do so elegantly and creatively.

If there is an agreed language tied to common constructs for what may be regarded as elegant and creative in Fish Creek,

Wisconsin, then Wisconsin's mathematical power will be of a similar quality to that described in the same terms at Pyramid Lake, Nevada, and Lodestone, California.

Therefore, in addition to the intellectual underpinnings, practical and logistical strategies are needed for arriving at interjudge agreement on standards of work between teachers within a school and between schools in different districts. Acceptable interjudge agreement will be achieved only through teachers' membership in a statewide community, collegial meetings within a school, and regional meetings between schools and between districts to arrive at and calibrate judgments. These structures exist in most states, but in such fields as athletics and music rather than in mathematics and language.

Vermont, for example, has not previously had state assessment. It seeks now to design a state mathematics assessment based on a combination of uniform tests, student portfolios, and student surveys. Its uniform assessment will enable NAEP comparison. Vermont's portfolio assessment in mathematics is intended to result in the assessment of more authentic mathematical activity. However, accomplishing portfolio assessment in mathematics requires substantial intellectual and practical preparation.

A committee of Vermont teachers has proposed that, for portfolio assessment in mathematics, teachers collect student work in folders. They decided to design portfolio assessment on the basis of assessing individual work and aggregating from that basis. For program assessment, entire portfolios of individual students would be assessed by an external team on a small sampling basis. For student assessment, the best three work examples of each student—selected by teacher and student—would be submitted for external review.

The committee solicited examples of student work and met to consider how each might be evaluated and how the resulting data might be used for state assessment. Student work that could be assessed in the context of a uniform test was set aside, and the efforts that drew the favorable attention of committee members were considered more closely. In the process, several things became apparent:

- there was a generally agreed upon, but poorly articulated, perception of the quality of mathematical activity among the group;

- as group members worked to spell out to each other what they were seeking as evidence of mathematical power, a series of rubrics and common standards of judgment emerged;

- even the best student work, including that of their own students, did not meet all the standards of excellence that members of the group were applying;

- typical worksheets were judged of little value in the context of portfolio assessment and were to be excluded;

- much student work had been brought to an end just as it had the potential to become mathematically interesting;

- the most interesting student work emerged in the context of such activities as design and explanation;

- articulate language and variety in representation by the students was, for those judging, a critical entry into the quality of their mathematical activity;

- introductory paragraphs from the teacher and the student would add significantly to the meaning of the work and the ability of an external assessor to evaluate it;

- some aspects of mathematical power, such as confidence, could be assessed only by the classroom teacher; and

- individual teachers planned modifications of their own work with students to bring it more in line with the standards for judging mathematical power that they had been helping to articulate.

The committee's initial draft of a coding scheme was based on an implicit emphasis on problem solving and an explicit concern with the five goals of the NCTM (1989) *Standards:* becoming a mathematical problem solver; learning to reason mathematically; learning to communicate mathematically; learning to value mathematics; and developing confidence in one's own ability to do mathematics. Vermont has made a powerful start in the direction of collecting direct evidence of authentic mathematical activity. The quality of evidence sought and the epistemological stance implicit in the proposed assessment structure are likely to spur real change. (See excerpts from the Report of Vermont's Portfolio Assessment Program in Appendix G.)

SUMMARY AND CONCLUSION: A RECOMMENDATION FOR CAP

The objective of this paper has been to propose reporting categories for the CAP that would simultaneously monitor and promote reform in mathematics education. The object of reform has been defined as the attainment of mathematical power. This is a complex and multifaceted view of mathematical literacy that requires a different epistemology for school mathematics.

A number of bases for forming reporting categories were outlined. Each reflects significant aspects of information about students' mathematical achievement and has advantages and disadvantages. Direct evidence of mathematical power requires categories that focus on the active use, generation, and communication of mathematical ideas in problematic situations and in collaborative contexts that are based in a familiar culture.

However, the sine qua non for reform is change in school beliefs about the nature of authentic mathematical activity and the character of the mathematical enterprise. Unless students have experience in generating mathematical ideas—seeing mathematics as part of their culture and becoming encultured into the mathematical enterprise (Bishop, 1988)—little of substance will have changed. The set of categories recommended has these two changes as its cohering purpose.

Nevertheless, the most appropriate set of reporting categories is still corruptible by other facets of the assessment system. It is clear that to gather the kinds of direct evidence regarded as essential will require data collection strategies different from those now in place for CAP or for any other general assessment program. A representative range of potential strategies has been succinctly and effectively summarized in *Assessment Alternatives in Mathematics: An Overview of Assessment Techniques for the Future* (Stenmark, 1989).

It is obviously impossible for CAP to use such measures for external assessment without creating a massive and expensive assessment bureaucracy. Such a strategy would be undesirable because it would be antithetical to the kind of epistemological change that has been advocated. The logical strategy is for CAP itself to re-empower teachers by involving schools and teachers in a collaborative assessment process that includes an element of student self-assessment. This would be a major step toward initiating epistemological change. It would require training and inservice for teachers, as well as common templates for task analysis, a common language of description, and formal organizational structures for developing and maintaining interjudge agreement.

The implication is that the California Assessment Program needs to develop a new character and a new program. To guide schools and teachers in the assessment process, it needs to:

- set standards for school-based assessment;

- train teachers to gather evidence;

- provide a structure for developing interjudge agreement;

- provide quality control over the process;

- act as a mentoring and moderating authority to initiate, sponsor, and adjudicate; and

- collect, analyze, and disseminate a much more complex range of information.

Assessments that incorporate student self-assessment, teacher involvement in the assessment process, alternative strategies for gathering information, and other such efforts are possible and being used in some places. Furthermore, the kinds of categories, the concerns discussed, and the solutions suggested have been independently arrived at in disciplines other than school mathematics. There is a converging view that the kinds of categories proposed—authentic activity in the domain, collaborative activity, facility in communication, and enculturation into the domain—have broad significance for authentic assessment. One cannot have both total freedom and total control, and democratic strategies are as essential to change in the epistemology of school mathematics as they are to the national economy.

13

Evaluation—Some Other Perspectives

Philip C. Clarkson

"Yes, but who will change the tests?"
(National Council of Teachers of Mathematics, 1989, p. 189)

It is apparent that "the tests" referred to are not the tests teachers give in their classrooms on a day-by-day or weekly basis. They already have control over these. "The tests" are the standardized assessment instruments which are used throughout the United States, often authorized by legislation, devised by commercial organizations, and seen by many teachers throughout the country as being a forceful factor in structuring the mathematics curriculum. The preceding papers in this volume introduce the issue of current testing practice into the ongoing debate and ferment that surrounds mathematics today. This chapter sketches developments in the State of Victoria, Australia, over the last 25 years where there is only one external test given at the end of the school system, in Year 12. This contrasting situation may contribute constructively to the ongoing debate in both Australia and the United States as to how to monitor the work of schools.

EVALUATION AND ASSESSMENT—GENERAL THEMES

The introductory paper and the two following papers tend to "set the scene" for the rest of the volume. The opening paper provides an overview, and the second paper a historical perspective in which to place the looked-for changes in assessment. It is well to remember that what is being undertaken today is not new in a fundamental way. Today's changes are the latest in an ongoing process. The third of these papers focuses discussion on the 1989 NCTM (National Council of Teachers of Mathematics) *Standards*. Within these three papers are found most of the themes on which other contributors to the volume elaborate.

One of these themes has to do with the fact that standardized tests have gained an undue influence over the curriculum. It has become evident that no matter what is in the syllabus, teachers will teach to what is examined. There are two points to be made about this: first, since by and large these tests examine skills and knowledge, school mathematics has reflected this fact in its emphasis on teaching skills and knowledge. It is noted that such an emphasis is quite at variance with the proposed changes in the curriculum. The second point stresses the fact that since teachers are very conscious of "what is examined," a change in the assessment procedures to emphasize the new goals will encourage teachers to change as well. This circularity has been summed up in the now familiar phrase, "What is tested is what gets taught" (Mathematical Sciences Education Board, 1989, p. 69). Whether standardized tests can be changed in such a way as to accurately reflect the changes proposed for the curriculum is examined; on balance, it seems doubtful whether they can be. Hence, the specific role that systems-wide testing may have in the changed educational environment will need to be examined explicitly.

Another theme that emerges in this volume follows from the first. The fundamental view of knowledge embodied in the present standardized testing procedure is the notion that knowledge, in this case mathematical knowledge, is out there waiting for students to consume it. The role of the teacher is to serve up this knowledge of mathematics in such a manner that the students will actually ingest it and assimilate it appropriately. It follows, in this analogy, that the role of assessment is to cue students to regurgitate their knowledge, often in forms which are only slightly digested. It is recognized that few students fully digest such knowledge, at least according to what the tests tell us. And it is very difficult in any case to unscramble digested knowledge reliably when you only use multiple-choice items. So unscrambling" is not often attempted, and because of the status that standardized testing has gained, any attempt at doing so is devalued.

But this volume argues that a far more rewarding way of thinking about knowledge in the educational context is to think of it as a process. It is a process by which students construct their own meanings for this area in their lives called mathematics. If this is so, then assessment needs to be thought of also as a process that provides some indication of the meaning students have accorded mathematics in their lives. It is like taking snapshots of a moving target. Perhaps different types of cameras positioned in strategic locations will be necessary to create an overall picture of what students know. Maybe it also calls for full investment on the part of the person most closely involved with them in the process, the teachers, and even other persons close to them, such as parents and peers. This approach is not compatible with that implied by standardized tests, which are supposed to provide an objective, clinical, scientific assessment of what students know, but more often tend to indicate what they do not know.

Following naturally from the above, the third theme emerges: this is the declaration that standardized tests cannot serve as appropriate assessment instruments for the collection of information of interest to diverse parties. And yet it is these scores that are used to tell teachers how their classes are going, to tell bureaucrats how specific schools are doing, to tell politicians how districts are doing, and to tell the nation how its education system is doing. In some areas, the results from these tests are also applied to individual students; hence, they and others have an interest in finding out how they are doing as well. The indefinite terms, "doing" and "going," have been used advisedly. Each of these different groups respond to essentially the same set of data, massaged in slightly different ways, to be sure. However, the meaning for each group is quite different. Bearing in mind that a specific objective should be articulated for any assessment instrument, it is hard to believe that one standardized test can be used with confidence to respond to the wide range of interest represented by the constituencies named above.

The last theme to deserve comment involves specifically one of the interested groups named in the last paragraph. Increasingly, the government and, particularly, politicians are demanding accountability for the many dollars invested in education.

There seems to have grown up in both the United States and Australia a management/economics style of government modeled in some ways on big business. Hence, virtually everything has to be accounted for in monetary terms. It is construed as a mark of responsible government for bureaucrats to demand that systems that receive federal funding, such as education systems, act for the public good. However, governments have often found that obtaining accountability reports that are meaningful is not always easy. As a case in point, the teaching/learning process in many ways is "messy," and in certain respects not easily reduced to simple terms. Like any process in action, it is often hard to capture its diversity and essential qualities at the same time in a simple account. But governments more often than not prefer simple accounts, and if the targeted data can be reduced to numbers, all the better. Numbers can be manipulated in many ways, and their use conveys the impression of objective, even scientific reporting. The results of extensive, mandatory tests comprise one such set of figures. But these, with other sets of figures, do not capture the real story of what is happening in schools and, therefore, in the system as a whole. The politicians are selling themselves short. There are good stories to tell and some that are not so good, but the telling is more ambiguous and complex than any set of figures can convey. This theme needs to be addressed more directly in any ongoing discussion of the means by which individual groups both within the system and outside it can communicate effectively with each other.

SPECIFIC ISSUES ASSOCIATED WITH
MATHEMATICS ASSESSMENT

In the above section, I have attempted to cite some of the general themes in this volume that are of greatest importance. However, there are also a number of specific issues that add other perspectives to the debate on evaluation. A few of these are summarized below.

The Use of Calculators and Microcomputers in the Classroom

One of these issues is the diverse response within the mathematics education community to the role of calculators and their place in assessment procedures. Electronic calculators have been with us for more than twenty years. Their incorporation into the curriculum, although called for sometime ago, is only really happening now. But the use of calculators in assessment procedures is still not a fact of life. Without question, there is still a residual resistance to the use of this technology in the classroom. Perhaps the reorientation to doing mathematics argued for in the *Standards* has not yet been accepted by many teachers.

Perhaps these factors, among others, play a role in the reluctance to incorporate the use of calculators into assessment procedures. At least, they raise interesting questions: If there had been a strong push for calculator use in assessment procedures early on, would this have led to an greater acceptance of them in the classroom? remembering that "What is tested is what gets taught." In turn, would this have led to different types of standardized tests being produced? Or has it partly been the ingrained place of standardized tests in the system and the nonuse of calculators in the tests that meant that this technology was regarded in a neutral or negative light by teachers? Furthermore, the corporations that produce such tests may even have regarded calculators as a threat. Given the extensive descriptions of test-item production provided in this volume, how much control do teachers and others at the district level have over standardized tests? These questions are worth considering.

Interestingly, in this volume there is no treatment of the role of the microcomputer in mathematics assessment. Perhaps with so much specific attention directed to standardized testing, this issue was less compelling. One hopes that if there is no role currently accorded micros in system-wide testing, this will not in turn devalue the microcomputer. That clearly would be at odds with the general sentiment of contributors to this volume regarding the use of technology in the teaching of mathematics. Rather than take that road, which may have been the one pursued in relation to calculators for too long, it may be better to question the place of standardized testing and its justification. But more of that later.

Problem Solving and Assessment

While it seems that calculators have not been accepted by the developers of standardized tests, the term *problem solving* has been. But an examination both of the papers in this volume that describe the production of these tests and those papers that question their past and present use clearly shows that the term is being used in a way different from that directed to the teachers by the NCTM. In the *Standards*, problem solving is seen as an all-pervading approach to mathematics. It is regarded as one way of doing mathematics and not simply as one cognitive level compared to, say, analysis. Yet the test developers have seen the term *problem solving* as another such category (leaving aside the question of the very use of categories canvassed extensively in this volume). Is this use of the term *problem solving* a cynical effort to dress the old tests in new terminology to make them acceptable in the new environment? Or are we witnessing a real shift, an initial attempt by the testing industry to respond to new directions? It is easy to believe the former in light of the analyses presented in these chapters, but perhaps the question needs closer examination. The one exception, the one state program approach which has responded to the call and uses *problem solving* in a manner compatible with that of the *Standards*, is the development of assessment in California.

Gender Bias in the Mathematics Classroom and in Assessment

Another specific area examined in this volume is that of gender differences in performance on tests composed of multiple-choice items. Since these types of tests predominate in standardized testing, the conclusion that such items may in certain cases favor males gives us pause with respect to the use of such instruments. This issue, too, encourages the use of other forms of assessment. It is suggested in a number of papers that a *variety* of assessment methods be employed. Indeed, using the analogy of a series of cameras positioned in different locations, the same point was made in the previous section of this paper. However, if there is concern about gender effects in multiple-choice items, there is clearly a need to investigate whether other forms of assessment are prone to the same type of problem.

One aspect of mathematics which is promoted in the *Standards* is that of communication. Certainly within the verbal discourse which goes on in the classroom at present there are gender differences. It has been known for a long time that teachers react to males and females in different ways. In a recent paper dealing with this issue, Leder (1990) noted that in Grades 3, 6, 7, and 10, Australian males interacted more frequently with teachers than females and tended to dominate the attention of the teacher. However, she noted that the differences in interactions were subtle. For example, although there were the same number of questions asked of males and females by the teachers, the type of response made by the teachers differed relative to the gender of the student in Grades 6, 7, and 10. In these grades, there was an indication that the teacher waited longer for answers from females on low cognitive questions, but for males they waited longer when the question was classified as high cognitive. Leder went on to suggest that these and other differences she found may well be signaling the students that there are differences in the mathematics they are supposed to construct that are dictated by their gender. Since both questioning and the model of questioning that the teacher employs are essential aspects of mathematics, as well as a fundamental aspect of the teacher's assessment strategy, any gender bias needs to be recognized for what it is.

Another aspect of communication promoted in the *Standards* is for students to write about their mathematics. Again, there may well be gender differences pervading this activity which the mathematics teacher needs to be aware of. Perhaps an examination of writing in language classes would be a useful place to start in investigating this, bearing in mind that there is no guarantee that results in such contexts will transfer exactly to a mathematical context.

Other Forms of Bias in Mathematics Assessment

Some further potentials for the examination of bias in assessment procedures may be those of language and culture. These have not been examined in this volume but need also to be addressed. There is the whole aspect of dealing with Mathematical English in monolingual classrooms and how that impinges on assessment procedures (see, for example, Newman, 1983, and Watson, 1980). This will clearly overlap with aspects of the gender bias issue noted above. However, it is also quite evident that in a significant number of classrooms in the United States (see Secada, 1990) and, for that matter, in other pseudo-monolingual countries such as Australia and England, there are many students whose mother tongue is not English and who are members of a minority culture. There has been some research on assessment and bilingual students (Cuevas, 1984), but the tests were of traditional types and were not conclusive. There has also been recent comment on students' cultural backgrounds and on the different styles of learning environments for mathematics (see Clarkson, 1991). It may be that for some cultures, using small groups might prove to be a decided advantage, but for others a disadvantage. If, however, assessment procedures for problem-solving situations envisage the use of cooperative small group work, then it may be important to look at the implications of bilingualism and the multicultural environment in which such groups may operate.

Other Issues

If large-scale testing is still to have a role to play in American education systems, then Mark Wilson's paper in this volume may prove valuable. It will certainly enable tests to be developed that are alternatives to the instruments of today. It is also of interest to note that work on linking the SOLO model with the work of van Hiele is already underway (Pegg & Davey, 1989). The new types of descriptors examined in Wilson's paper would especially be of use. However, these descriptors have the added advantage of being useful to teachers as well, and hence empowering the process of teaching. They clearly imply that more than one type of assessment procedure must be used for a clear picture to emerge of what happens during the teaching and learning of mathematics in the classroom. The brief report of the Australian research projects extends this idea.

The summary of results from journal writing developed in Victoria is perhaps more than just one other example of how teachers can gain insight into the way their students develop mathematical ideas. It is more too than a student's own record of self-dialogue. In some of these results, there is an indication that the change called for in the *Standards* may be attainable. The results suggest that teachers may well be more interested in the strategies that students use in problem solving than in whether they have acquired the set knowledge. The reports that students can devise their own maps of knowledge serve as an interesting parallel to the call for teachers to do just that. The data that suggest students can distinguish between the difficulty of a problem and the type of thinking required to solve it are also of interest. These last observations in combination with other data lead to the conclusion that if students are given appropriate techniques and an environment conducive to learning, they can indeed assume control of their own learning process. This set of data also clearly shows that feelings — on the part of students and teachers— are part of the whole learning/ teaching/assessment process. To ignore them is to ignore an important element in that process. How do we recognize feelings and their role in this new approach?

SOME PERSONAL NOTES ON ASSESSMENT

The papers in this volume discuss current assessment procedures in mathematics, including the use of standardized testing in the United States, and suggest a number of options for consideration as the impact of change in mathematics curriculums is felt in the classroom. However, there has been little attempt to open up directly the question of whether systems-wide testing procedures in mathematics should continue to be employed in the U.S. now that the new curriculum changes are taking hold. There have certainly been some implicit suggestions that such testing will need to change radically; the feeling is that such testing mechanisms will not be useful in the future. Change has also been occurring elsewhere. It would be illustrative to sketch briefly an example of change in schools where there has been little use of systems-wide mandatory testing for many years. The contrasting situation may provide another perspective to the ongoing debate.

Change in Mathematics Assessment in Victoria's Schools

Toward the close of the 1950s in Victoria, there was mandatory assessment at the end of Years 10, 11, and 12. Prior to this time, there had been examinations in earlier grades as well, but they had been dispensed with. By the early 1960s, most schools were even afforded the privilege of assessing Year 10 students internally with no recourse to external, education-department-approved tests. Indeed, some larger schools were accorded the privilege of assessing Year 11 students internally. Schools were accredited on the strength of how well qualified their staffs were and who had experience teaching the subjects in question. Part of this accreditation was based on the strength of reports made by departmental inspectors who visited the schools and classrooms on a regular basis.

The curriculum that was used in schools at that time reflected the expectations of the system. The system was like a tunnel. Students entered at one end and it was presumed that they would progress through the school experience until at the other end of the tunnel they made the transition to the university level. Melbourne University, the only university in Victoria at that time, had a great deal of influence on the school curriculum, viewing it primarily as a private preserve. The reality, of course, was very different. Most children were pushed out of the system, or left of their own accord, before Year 11. However, for many years the curriculum, based on the traditional schooling system inherited from the English, simply did not reflect this situation.

There were three mathematics subjects each in Years 10, 11, and 12. In Year 12, the subjects were designated Pure, Applied, and General Mathematics. Students wishing to take tertiary courses which required mathematics took the Pure and Applied level courses. General Mathematics was considered an easier option and was not recognized as a prerequisite for future study. It had been introduced as a way for returned servicemen from the war to meet the university's entry requirement for first year students, which mandated that they pass a modern language or a mathematics test at Year 12 level. However, in succeeding years, some faculties at Melbourne University did recognize General Mathematics in fulfillment of their prerequisite requirements. Each mathematics examination was composed of about ten extended items. Students were advised that complete answers to about seven would be worth full marks.

The year 1966 was important in Australian education. At that time new courses of study for Years 10, 11, and 12 were issued by a new board of the department of education. Suggested courses of study for Years 7 through 9 were also included, and these, for all intents and purposes, became the official syllabi. This board was composed of representatives of the education department, teachers, Melbourne University faculty, and, importantly, faculty of the new Monash University, among others. The singular influence of Melbourne University was now challenged. The new board proposed two syllabi for each mathematics subject: a school could choose to teach a variation of the traditional syllabus, or a new one based on the New Mathematics. By 1972, it was clear that the "new" syllabus would become the standard for all schools. One of the interesting points to note is the freedom given to schools to make the choice individually as to which syllabus to follow in the intervening period. The examination associated with this new syllabus was different as well. All components were of an extended item type. These were divided into two sections; the first containing items worth up to three marks each and the second section containing items worth up to ten marks or so. In 1966, the education department also issued its last syllabus for primary schools.

In 1972, the new mathematics syllabus became the sole standard. That year also saw the phasing out of Year 11 assessment requirements. This meant that Year 12 was the only year in which mandated assessment was required of students completing their secondary schooling. This examination was still heavily influenced by the tertiary education sector, other colleges also being represented on the examining board by then. However, while most students still left school before reaching Year 11, the school curriculum was just beginning to reflect this fact. Teachers were starting to take greater control of their teaching; and they were starting to teach their students, rather than being cowed into teaching a curriculum solely to prepare students for university courses—even though many would not, and never intended to, go to the university. Interestingly, it was in 1972 that the education department issued for primary schools a *suggested* course of study in mathematics rather than an official syllabus.

There were at this time in Victoria other education activities that also impacted on the schools. Throughout the 1960s, for example, the teachers' unions were asking the increasingly strident question: If teachers are professionals, why are they scrutinized at regular intervals? During the 1970s, first in the secondary schools (Years 7 through 12) and later in the primary schools (Years Kindergarten through 6), the education department withdrew its inspectors as observers of classroom teachers and, finally, withdrew them from the schools altogether. The curriculum continued to change and teachers took advantage of their newfound freedom. In mathematics, teacher groups like the Study Group for Mathematics Learning set out to encourage secondary teachers to explore new ways of approaching the teaching of mathematics, such as using supplementary materials in their teaching. Other groups, like the Rusden Activity Mathematics Project set out to develop written materials for teachers to use based upon activities in the classroom. They eventually published twelve booklets for use in Years 6 through 8. Another group, the Mathematics Association of Victoria, became the nucleus of mass-based teacher organizations that developed and conducted many forward-looking workshops and conferences by and for teachers. Other educational developments of broad significance included the regionalization of the once extremely centralized department of education.

By the mid-1970s, the education department had re-formed its examination board and included representatives of employer groups as well as a wide array of education groups. The new board was not just to take charge of the Year 12 syllabus, but it was also to take an interest in the whole of the secondary sector of education. By the end of the decade, this board had revamped the Year 12 examinations. In mathematics, there were still the three subjects that had been offered since the 1950s; however, within each subject there was a designated core of study comprising about two-thirds of the content and then a number of options from which a school could choose to teach. One of these options dealt with computers in mathematics. The examination given at the end of Year 12 was no longer the only assessment tool used. The teacher was authorized to allocate a score for the study of the optional section. There were various coordinating devices used to help ensure compatible marks for students from different schools. Teachers attended meetings throughout the year during which the course was discussed. There was also a process of statistical comparison during which the internally allocated scores were adjusted to the mean and the standard deviation of the external scores obtained by a particular school.

The re-formed board also recognized other subjects. The three traditional subjects were designated Group 1 subjects. There were also Group 2 subjects, which included no element of external assessment in their curriculums. All assessment was carried out within a school with various comparison strategies employed between groups of schools. An accreditation body was formed to oversee the operation of these subjects. In the mathematics area, Group 2 subjects included Mathematics for Work and Business Mathematics, among others. Not all universities recognized these subjects, but some other tertiary colleges and many employers did. A number of other certificates were also recognized along with the traditional end-of-schooling certificate. One certificate was based on the philosophy that the curriculum should be negotiated between teacher and student. An accrediting body was constituted by the education department which established broad guidelines on content and on procedures for comparative evaluation. Although not many schools incorporated this option into their curriculums, those that that did reported great success with it. It will be appreciated that these alternative curriculum styles fostered experimentation with a number of different assessment styles.

In the mid-1980s, a revamping of the mathematics subjects in Year 12 finally resulted in the end of General Mathematics. This subject had been seen as the soft option by students for many years. Since the mid-1970s, mathematics teachers had been trying to have it deleted against the opposition of a number of university faculties. Another change was that calculators were expected to be used when completing the Year 12 examinations. These and other revisions of the curriculum were the beginning of a wider move by the department of education to update the curriculum in Years 11 and 12. Also discarded were the alternative Group 2 subjects—a move many teachers regarded as unfortunate. Among other non-curriculum changes instituted in the decade of the 1980s was the introduction of self-management in individual schools.

The New Mathematics Curriculum

A set of "new" mathematics units will be taught in all Victorian schools for the first time in 1991 in Year 11 and in 1992 in Year 12. The department has again stipulated a course for Year 11, the first time this has been done for twenty years. This has resulted in a certain amount of opposition from teachers who believe they are losing some control. However, the semester unit structure that replaces the year long courses brings with it a lot more flexibility. The assessment procedures are of particular interest. In brief, there will be two categories of assessment:

the first deals with "completion of the unit," while the second focuses on the "level of achievement." For each semester unit there are specific tasks which a student needs to finish in order for the subject teacher to report on the certificate that the unit was "successfully completed." This procedure is followed for both years. The teacher and school are also responsible for judging the "level of achievement" if a unit is taken at the old Year 11 level. The assessment techniques employed are at the discretion of the teacher and the school. However, for equivalent Year 12 units, there are four common assessment tasks which must be used. At the time of writing, it is believed that one of these tasks will be a $1^{1}/_{2}$-hour examination that is external to the schools and composed of fifty multiple-choice items. This examination will be mainly aimed at skills.

The second task will be another externally set examination but one composed of extended-answer items designed to examine higher-level skills. Both of these examinations will be marked by external examiners. The use of calculators will continue to be expected when completing papers. The third task will be a project which will involve an extended writing assignment. A problem-solving task will be the fourth. These last two tasks will be evaluated by the subject teacher and then submitted to a process of comparison with other teachers and moderators before final marks are arrived at. These tasks could be completed individually, but there is scope in the procedures for group work as well. Indeed, it is hoped that small group work will be a common approach. Separate lists of problems for each of these tasks will be circulated, and two- to four-week time slots will be designated during which the tasks will have to be completed. Students will select the problems they wish to work on. The use of microcomputers is to be encouraged. Finally, in the reporting process there will be no attempt at combining the four resulting scores into a global score for a mathematics unit. The four letter grades per unit attempted are to be reported separately as letter grades on the certificate that the student receives (Victoria Curriculum and Assessment Board, 1989).

This offers a general outline of how senior school mathematics has changed in thirty years in Victoria. From being dominated by the one university in the state via external examinations, the students now receive a certificate that indicates their achievements at the end of secondary school. The curriculum is not completely dominated by what is expected in the first year of university study. There is now recognition that only some students will choose to go to university as the next step after school. Teachers, reinforced by ex-teachers, have exerted great influence over the curriculum they teach, and indeed over the schools, not to the exclusion of outside interests, but in a balanced way. They have been expected to act as professionals. And they have done so.

During this process, it was recognized that the only acceptable point for control of an examination by those outside a given school was in Year 12. That calculators as well as microcomputers can be used when completing assessable tasks has been accepted. A way has been found to include a range of different types of tasks on which to judge a student's mathematical knowledge, and there is some room for the student's own choice of problem.

Victoria's system is not perfect. There are certain participants in the process who are not satisfied. Among these are the tertiary institutions. Since they have used a combination of students' Year 12 final marks as an entry score because of the ease with which such numbers can be computed (even though it has been acknowledged as an illogical computation), they are not happy with up to four letter grades on different tasks for each of up to twenty-four different units (both mathematical and nonmathematical units).

It is acknowledged that this process of change will not stop here. Change will continue, and the subsequent changes will undoubtedly be built on present experience. There have been a variety of programs offering alternative assessments for a number of years in Victoria. The present full scale implementation has drawn from many of them. The use of calculators was a gradual process in assessment procedures until fully implemented in the early 1980s. The present situation represents a point reached after many years of change.

Nor is the situation in Victoria a blueprint for any other state, region, or province. The quite different pressures and circumstances in each locality prevent this. However, this summary is offered as an example of what can be done; it is not perfect by any means, but a stimulating example perhaps for others.

SUMMARY NOTES

Perhaps the major thrust of the changes in mathematics looked for and epitomized by the *Standards* document is the need to empower the teacher. After all is said about systems-wide testing programs, the overriding feeling is they are predominately used to check on teaching quality. And in one sense, such a conclusion is correct. Perhaps the crucial factor in schooling is teacher quality. But to use student tests to judge teaching quality is to employ a rather indirect method. The teachers are certainly aware of the reason these tests are given and respond accordingly. However, their response is not positive, but rather one which prevents them from reacting in creative ways to the situations that arise in their own classrooms. A quote from the penultimate paper in this volume seems to sum up the point:

> If one really subscribes to the idea that a change in the authority structure of school mathematics is essential to real change in its epistomology, that authority must be seen to transfer from external experts to the school, the teacher, and the student. The assessment process, including its tasks, is a key part of that process. As long as assessment is entirely an external dictate rather than a collaborative effort, the final answer for students is that they must learn somebody else's mathematics as opposed to holding their own mathematical ideas up for cooperative assessment by an entire mathematical community, which includes their peers (Zarinnia & Romberg, this volume, p. 275).

APPENDIX A

NCTM EVALUATION STANDARDS

General Assessment

Standard 1. Alignment
In assessing students' learning, assessment methods and tasks should be aligned with the curriculum in terms of:

- its goals, objectives, and mathematical content;
- the relative emphases it gives to various topics and processes and their relationships;
- its instructional approaches and activities, including the use of calculators, computers, and manipulatives.

Standard 2. Multiple Sources of Information
Decisions concerning students' learning should be based on the convergence of information obtained from a variety of sources. These sources should embody tasks that:

- demand different kinds of mathematical thinking;
- present the same mathematical concept or procedure in different contexts, formats, and problem situations.

Standard 3. Appropriate Assessment Methods and Uses
Assessment methods and instruments should be selected on the basis of:

- the type of information sought;
- the use to which the information will be put;
- the developmental level and maturity of the student.

Use of assessment data for purposes other than those intended is inappropriate.

Note: From the "Overview of the Curriculum and Evaluation Standards for School Mathematics" (An Abridgement and Excerpts of NCTM's *Curriculum and Evaluation Standards for School Mathematics*). Prepared by the Working Groups of the Commission on Standards for School Mathematics, NCTM, October, 1988. pp. 16–18.

Student Assessment

Standard 4. Mathematical Power
The assessment of students' mathematical knowledge should seek information about their:

- ability to apply their knowledge to solve problems within mathematics and in other disciplines;

- ability to use mathematical language to communicate ideas;

- ability to reason and analyze;

- knowledge and understanding of concepts and procedures;

- disposition towards mathematics;

- understanding of the nature of mathematics; and integration of these aspects of mathematical knowledge.

Standard 5. Problem Solving
The assessment of students' ability to solve problems should provide evidence that they can:

- formulate problems;

- apply a variety of strategies to solve problems;

- solve problems;

- verify and interpret results;

- generalize solutions.

Standard 6. Communication
Assessment of students' ability to communicate mathematics should provide evidence that they can:

- express mathematical ideas by speaking, writing, demonstrating, and depicting them visually;

- understand, interpret, and evaluate mathematical ideas that are presented in written, oral, or visual forms;

- use mathematical vocabulary, notation, and structure to represent ideas, describe relationships, and model situations.

Standard 7. Reasoning

The assessment of students' ability to reason mathematically should provide evidence that they can:

- use inductive reasoning to recognize patterns and form conjectures;

- use reasoning to develop plausible arguments for mathematical statements;

- use proportional and spatial reasoning to solve problems;

- use deductive reasoning to verify conclusions, judge the validity of arguments, and construct valid arguments;

- analyze situations to determine common properties and structures;

- appreciate the axiomatic nature of mathematics.

Standard 8. Mathematical Concepts

Assessment of students' knowledge and understanding of mathematical concepts should provide evidence that they can:

- label, verbalize, and define concepts;

- identify and generate examples and nonexamples;

- use models, diagrams, and symbols to represent concepts;

- translate from one mode of representation to another;

- recognize the various meanings and interpretations of concepts;

- identify properties of a given concept and recognize conditions that determine a particular concept;

- compare and contrast concepts with other related concepts.

In addition, assessment should provide evidence of the extent to which students have integrated their knowledge of various concepts.

Standard 9. Mathematical Procedures

The assessment of students' knowledge of procedures should provide evidence that they can:

- recognize when it is appropriate to use a procedure;

- give reasons for the steps in a procedure;

- reliably and efficiently execute procedures;

- verify results of procedures empirically (e.g., using models) or analytically;

- recognize correct and incorrect procedures;

- generate new procedures and extend or modify familiar ones;

- appreciate the nature and role of procedures in mathematics.

Standard 10. Mathematical Disposition

The assessment of students' mathematical disposition should seek information about their:

- confidence in using mathematics to solve problems, to communicate ideas, and to reason;

- flexibility in exploring mathematical ideas and trying alternative methods in solving problems;

- willingness to persevere at mathematical tasks;

- interest, curiosity, and inventiveness in doing mathematics;

- inclination to monitor and reflect upon their own thinking and performance;

- valuing of the application of mathematics to situations arising in other disciplines and everyday experiences;

- appreciation of the role of mathematics in our culture and its value as a tool and as a language.

Program Evaluation

Standard 11. Indicators for Program Evaluation

When evaluating a mathematics program's consistency with the NCTM *Standards,* indicators of the program's match to the *Standards* should be collected on:

- student outcomes;

- program expectations and support;

- equity for all students;

- curriculum review and change.

In addition, indicators of the program's match to the *Standards* should be collected on curriculum and instructional resources and instruction. These are discussed explicitly in Evaluation Standards 12 and 13.

Standard 12. Curriculum and Instructional Resources

When evaluating a mathematics program's consistency with the NCTM *Curriculum Standards,* examination of curricular and instructional resources should focus on:

- goals, objectives, and mathematical content;

- relative emphases on various topics and processes and their relationships;

- instructional approaches and activities;

- articulation across grades;

- assessment methods and instruments;

- availability of technological tools and support materials.

Standard 13. Instruction

When evaluating a mathematics program's consistency with the NCTM *Curriculum Standards,* instruction and the environment in which it takes place should be examined, with special attention to:

- mathematical content and its treatment;

- relative emphases assigned to various topics and processes and the relationships among them;

- opportunity to learn;

- instructional resources and classroom climate;

- assessment methods and instruments used;

- the articulation of instruction across grades.

Standard 14. Evaluation Team

Program evaluation should be planned and conducted with the involvement of:

- individuals with expertise and training in mathematics education;

- individuals with expertise and training in program evaluation;

- decision makers for the mathematics program;

- users of the information from the evaluation.

Note: From the "Overview of the Curriculum and Evaluation Standards for School Mathematics" (An Abridgement and Excerpts of NCTM's *Curriculum and Evaluation Standards for School Mathematics*). Prepared by the Working Groups of the Commission on Standards for School Mathematics, NCTM, October, 1988. pp. 16–18.

APPENDIX B

CLASSIFICATION MATRIX

TEST NAME: _____

NUMBER OF ITEMS FOR EACH CATEGORY

QUES	CONTENT						PROCESS						LEVEL	
1	nr	ns	alg	p/s	geo	mea	com	c/e	con	rea	ps	p&f	conc	proc
2														
3														
4														
5														
6														
7														
8														
9														
TOT.														

Key:
nr — Number and Number Relations
ns — Number Systems and Number Theory
alg — Algebra
p/s — Probability or Statistics
geo — Geometry
mea — Measurement
com — Communication
c/e — Computation or Estimation
con — Connections
rea — Reasoning
ps — Problem Solving
p&f — Patterns and Functions
conc — Concepts
proc — Procedures

APPENDIX C

TEST RESULTS: PERCENT OF ITEMS FOR EACH CATEGORY

TEST	CONTENT						PROCESS						LEVEL	
	nr	ns	alg	p/s	geo	mea	ps	com	rea	con	c/e	p&f	conc	proc
SRA	82	7	7	0	4	0	3	5	1	0	91	0	16	84
CAT	73	5	6	6	4	6	0	11	6	0	83	0	10	90
SAT	64	0	10	9	2	15	0	38	0	0	62	0	8	92
ITBS	62	11	7	3	4	13	0	9	1	0	89	1	4	96
MAT	66	6	0	5	8	15	0	21	0	0	79	0	12	88
CTBS	76	0	0	11	8	5	0	25	2	0	71	2	15	85
AVG.	71	3	5	6	5	9	1	20	1	0	79	1	11	89
RNG.	20	7	10	11	6	15	1	18	2	0	20	2	11	12

Key:
	nr	Number and Number Relations
	ns	Number Systems and Number Theory
	alg	Algebra
	p/s	Probability or Statistics
	geo	Geometry
	mea	Measurement
	ps	Problem Solving
	com	Communication
	rea	Reasoning
	con	Connections
	c/e	Computation or Estimation
	p&f	Patterns and Functions
	conc	Concepts
	proc	Procedures
	AVG	Average
	RNG	Range

APPENDIX D

ILLUSTRATIVE QUESTIONS

Example 1 (multiple choice): Typically students are asked to solve for *x* when given an equation. In the following question, students are required to see an algebraic representation as mathematization of a real-world problem.

> Which one of the following problems
> can be solved by using the equation
> x + 2 = 28?
>
> O A math class started with 28 students.
> The next day 2 more students enrolled
> in the class. How many students does
> this class have now?
>
> O Erin added 2 more books to her
> collection. If she now has 28 books,
> how many books did Erin have
> originally?
>
> O Tim had $28 in his account. A week
> later he deposited $2 more. How
> much money does he have in his
> account now?
>
> O Ann biked 28 km at 2 km per hour.
> How long did Ann bike?

Example 2 (multiple choice): This question can be done in several ways depending upon the mathematical sophistication of the student. It can be approached purely by trial and error or by trial and error in a systematic way using knowledge of place value.

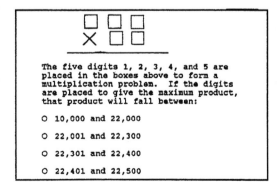

> The five digits 1, 2, 3, 4, and 5 are
> placed in the boxes above to form a
> multiplication problem. If the digits
> are placed to give the maximum product,
> that product will fall between:
>
> O 10,000 and 22,000
>
> O 22,001 and 22,300
>
> O 22,301 and 22,400
>
> O 22,401 and 22,500

Example 3

Name M___ A___   3-DEC-87   # 01416

Instructions: Use this sheet to answer the questions. Show as much of your work as possible. (In some cases, there may be more than one solution.) Use the reverse side of this sheet if needed.

Imagine you are talking to a student in your class on the telephone and want the student to draw some figures. The other student cannot see the figures. Write a set of directions so that the other student can draw the figures exactly as shown below.

a)

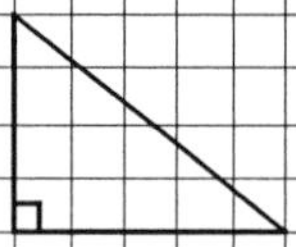

I WILL GIVE YOU DIRECTIONS USING A STANDARD CARTESIAN PLANE. THE BEGINING OF THE FIGURE WILL BE FROM (0,0).

START. DRAW A LINE FROM (0,0) TO (5,0). FROM (5,0) DRAW A LINE TO (0,4). FROM (0,4), DRAW A LINE TO (0,0).

END.

b)

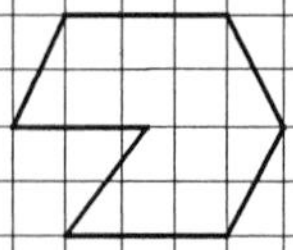

SAME DIRECTIONS AS ABOVE.

START. DRAW A LINE FROM (0,0) TO (3,0). DRAW A LINE FROM (3,0) TO (4,2). DRAW A LINE FROM (4,2) TO (3,4). DRAW A LINE FROM (3,4) TO (0,4). DRAW A LINE FROM (0,4) TO (-1,2). DRAW A LINE FROM (-1,2) TO (1½,2). DRAW A LINE FROM (1½,2 TO (0,0)

For Teacher Use Only

Example 4

California Assessment Program

OPEN-ENDED MATHEMATICS QUESTION(S)

Survey of Academic Skills

A

Name J_____ H _____

Instructions: Use this sheet to answer the questions. Show as much of your work as possible. (In some cases, there may be more than one solution.) Use the reverse side of this sheet if needed.

Imagine you are talking to a student in your class on the telephone and want the student to draw some figures. The other student cannot see the figures. Write a set of directions so that the other student can draw the figures exactly as shown below.

a) Draw a right triangle with a base of 5 units and a height of 4 units.

b) Draw the perimeter of a figure of an isosceles trapezoid sharing the lower right half of its base with a parallel side of an irregular trapezoid.

The base of the isosceles trapezoid that is not connected is 3 units long. The base of it that is attached to the parallelogram is 5 units long with a height of 2. The irregular trapezoid has a base of 2½ units, which is connected to the isosceles trapezoid, and a base of 3 units, which is placed 1 unit from the point where the lower vertex of the irregular trapezoid is attached with a height of 2.

For Teacher Use Only

APPENDIX E

HISTORY AND RATIONALE FOR STUDENT MATHEMATICS JOURNALS: A SCHOOL PERSPECTIVE.

A. Waywood, Mathematics Co-ordinator
Vaucluse College
Richmond, Victoria, Australia

Vaucluse College is a Catholic secondary girls school. There are approximately five hundred girls from Year 7 to Year 12 at Vaucluse. It serves a multicultural population: 20 percent Asian, 30 percent Italian and Greek, with the remaining 50 percent being predominantly Anglo-Saxon. Prior to the introduction of the mathematics journals, the mathematics program had been fairly "text book traditional." Mathematics was a compulsory part of the curriculum until Year 10. In Year 11 students could drop maths completely, do a Mathematics and Work unit, or continue with the core mathematics. Of students remaining at Vaucluse to complete their final year, about 30 percent would continue with mathematics.

History

In 1986, mathematics journals were introduced experimentally in one class each at Year 7, 9, and 10 levels. Compared to the present understanding of the functioning of a mathematics journal, these initial experiments were very crude in terms of the perceived relationship between students keeping a journal and a pedagogy of mathematics. Even so, results were encouraging enough to warrant the expansion of their use. By early 1989, the keeping of mathematics journals was seen as an essential element in the teaching of mathematics from Years 7 to 10. Years 11 and 12 need to be bracketed out of this discussion because, even though students are required to keep summary books, they are geared to an examination system where the skills developed through journal keeping are not prized. A student's completion of a mathematics journal should result in more than just a summary of mathematical procedures; rather it is the developing of an attitude to the doing of mathematics and coming to an understanding of what mathematics can be for students. In other words, in introducing journal keeping to the mathematics classroom, something has to happen for the student and the teacher.

Educational Rationale

Through 1987 and 1988, we worked hard trying to see how journals were functioning in terms of student learning and using these insights in a formulation of the purpose for having students keep a mathematics journal. This purpose is premised on the beliefs that language and thought are intimately connected and that mastering forms of communication goes hand in hand with mastering thinking.

By keeping a mathematics journal we intend that students:

1. Formulate, clarify, and relate concepts,
2. Appreciate how mathematics speaks about the world,
3. Think mathematically,
   a. Practice the processes (problem solving) that underlie the doing of mathematics,
   b. Formulate physical relations mathematically.

This purpose is translated into a number of tasks for students to do when they write a mathematics journal. Basically, any journal entry should be structured around three activities: summarizing, discussing, exemplifying. As a first introduction to journal writing, we supply our Year 7 students with an actual book, in which each page is divided into the sections: What we did, What I learned, Examples, and Questions. As an activity, each of these tasks has an internal structure that has the potential to draw the students into the experience of thinking systematically and of taking control of their learning. The actual structure of the tasks requires learning activity rather than learning passivity.

As I pointed out earlier, the journal activity has two roots: one in the dynamic of student learning, the other in the dynamic of instruction. Even though each of the tasks is essential for learning, none of them are explicitly taught in a mathematics classroom. We found that journals fulfilled our purpose to the degree that we accommodated our teaching to their use. Put most globally, using journals as an instructional tool required a shift from teaching techniques to helping students construct meaning. At the level of classroom implementation, this required teachers to:

- appropriate new models of mathematics instruction, such as group work, library research, historical investigations, and class discussion;

- experiment with non-traditional instructional devices, such as semantic maps, language of argument, modelling precise expression (formulating definitions), and redrafting.

To sum up, then, the educational rationale behind journal writing is to have students experience sustained and precise thinking. We suggest that language activity and mathematical activity come together and are focused in the act of precise articulation, which is the underlying demand of journal writing. Further, these two worlds of activity come together uniquely, because of the content of mathematics, which has to do with the relation between ideas and not about the relationship between things. Much of this rationale can be exemplified in a discussion of what journal completion entails.

What Journal Completion Entails

Students are required to write in their journal after every mathematics lesson. This is seen as ongoing homework. It is a requirement that is taken seriously because journals contribute 30 percent to the assessment in mathematics. As a minimum, a satisfactory journal entry should reflect the intellectual involvement of the student in the day's lesson. What form a particular entry will take is determined by the form of the day's lesson and the level of sophistication at which the student can interpret the journal tasks. To simplify this discussion, I will characterize lessons as falling into one of three types, Theory, Practice, Activity, and discuss under each the appropriate journal activity.

Journal Entry Appropriate to a Theory Lesson

The students will have taken notes in class and then that night will reconstruct the lesson and present a clear summary of the lesson. While doing this, they will note connections with previous ideas and concepts or applications that weren't clear to them. They will discuss what wasn't clear with the aim of phrasing a precise question that will get to the bottom of what they have not understood. The discussion will also aim to extend the ideas through the use of "What if . . . ?" questions. Where appropriate, they will give examples that illustrate the ideas or applications being discussed.

Journal Entry Appropriate to a Practice Lesson

After a practice lesson, students will spend time annotating a worked example. They will demonstrate an understanding of the connection between techniques and applications with the theory. They will isolate areas of background knowledge that prevent mastery of new techniques and test their understanding by doing a hard example. They will comment on their particular pattern of mistakes.

Journal Entry Appropriate to an Activity Lesson

In the first place, students will unearth the relationship of the activity to the unit of study, they will record what was done and discuss what it means, and they will reflect on how it illustrates ideas (for example, principles). They will describe and justify the method they have followed and state the conclusions they have reached.

It should be clear that the journal calls on many high-order processes. Being able to write a journal entry is not automatic for any student. Learning to use a journal has to be taught, and our experience is that if it is taught and applied during instruction in mathematics, then students find mathematics more meaningful, useful, and enduring.

What Constitutes a Successful Journal

A corollary to the issue of what journal teaching entails is the issue of how teachers recognize a successful journal. As we gained experience with reading journals, it became clear that something more was happening in journals that were seen as successful rather than just being complete. Successful journals were written differently. In the first instance, students who seemed to be getting the most from their journal work were more often trying to explain rather than just describe. From this insight, we formulated a taxonomy of the function of language in journals, which spread entries along a continuum. Students used language in a Narrative, Summary, or Dialogue mode. These terms are necessarily technical and are defined, at present, by examples of student work. What was most useful in these categories was that they gave teachers a means to discriminate between journals and to model proper use. Our contention was that as students learned to explain rather than describe, where summarizing was seen as a precursor to explaining, they were more likely to be thinking mathematically. This taxonomy of text has been very useful in judging successful journal completion and, further, seems to point towards differing dispositions of students towards mathematics.

APPENDIX F

SMP PROJECT: STUDENT LOG— SAMPLE PAGES

SMP 11–16 Coursework

London and East Anglian Group
Midland Examining Group
On behalf of Groups nationally

Candidate's Name	R_____

School	
Centre No.	Candidate No.

Brief title of task	Smarties

PLANNING SHEET FOR OPEN-ENDED TASKS　　Sheet [　]

Prices of packaging already in production.
Average amount of smarties in various packages.
Average price of smarties.
Price of cardboard packaging.

I needed to design a new box that is better value for money—i.e., less money for packaging and more smarties in it.

Draw the nets of boxes to find out how much cardboard is needed. Calculate amount of wastage. Assuming Rountrees have a 100% profit margin, calculate prices of boxes. Also prices of smarties.

New packages—designs which are not already on the market. Find out prices of plastic packaging.

SMP 11–16 Coursework

London and East Anglian Group
Midland Examining Group
On behalf of Groups nationally

Candidate's Name	R

School	
Centre No.	Candidate No.

Brief title of task	Design a new Smarties package.

LOG SHEET FOR OPEN-ENDED TASKS

Sheet [   ]

Date	Work Done	Queries	Teacher's Notes
Friday 25th Sep.	My father contacted Rowntree's in York to ask about prices of packaging. He was told the price of plastic tops but was not allowed the price of boxes. I was unable to ring myself because it was during school hours.		No information available about packaging prices or profit margins. Assumptions will have to be made if this idea is to be continued. WKM
Saturday 26th Sep.	Bought a bag of party packets.		
Sunday 27th Sep.	Found out average amount of smarties in party packets. Calculated amount of cardboard needed to make a cylinder. This was my first idea for a new package. Worked out the amount of plastic needed for the top.		
Monday 28th Sep.	I started to write up my notes and ...		

No wastage
fRom tube

Wastage fRom Box

Wastage fRom Party Packet

Bottom of tube

Party Packet

Wastage

Total cost (gross) = $\begin{array}{r} 0.59535\,p \\ 9.075\,p \\ \hline 9.67039\,p \end{array}$ +

Price × 100% profit = 19.34078
= 19p (to nearest pence)

Value for money (approx.)

Cylinder	Cube	Drum	Triangle
88p /259 smarties = 0.34p	26p /73 smarties = 0.30p	40p /113 smarties = 0.35p	19p /55 smarties = 0.35p

Which one is best?
The cylinder is the most economic design but
as smarties are mainly for children they wouldn't
want a packet quite so large.
The cube is a nice neat packet but is the
most expensive to make.
The drum is very good because it is economic and
it is quite a pleasing size for children.
The triangle is very economic and again it is
a pleasing size and also shape for children.

Taking all this into consideration I have chosen
Design 4. Although design 3 uses less cardboard, the
plastic puts up the price. 40p is probably a little
too dear for a packet of smarties. Although you
would sell many packets at this price I think more
packets would be sold at 19p. This packet is also
easily disposable whereas the plastic top of the drum is not.

Looking Beyond "The Answer"

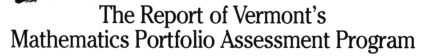

The Report of Vermont's Mathematics Portfolio Assessment Program

Pilot Year
1990-91

State of Vermont
Governor
Richard A. Snelling

State Board of Education

Douglas I. Tudhope, Chair .... North Hero
Sally Sugarman, Vice Chair .... Shaftsbury
Barbara M. Forest .... Castleton
Kathryn A. Piper .... Lower Waterford
Patrick S. Robins .... Burlington
Frances Elwell .... Brattleboro
Ross Anderson .... Burlington

The Vermont Department of Education
Commissioner of Education
Richard P. Mills

Director of Planning and Policy Development
W. Ross Brewer

Mathematics Consultant
Robert Kenney

TBA Consulting Group
William J. Thompson

II. Problem Solving and Communication: The Criteria

The best pieces of student work provide the basis for assessing the problem solving and mathematical communication skills of Vermont students. This section provides a description of these two elements and the seven criteria by which these are measured.

Problem-Solving Skills
  PS1. Understanding of the Task
  PS2. Selection of Approaches/Procedures/Strategies
  PS3. Use of Reflection, Justification, Analysis, Verification in Problem Solving
  PS4. Findings, Conclusions, Observations, Connections, Generalizations
Mathematical Communication Skills
  MC1. Language of Mathematics
  MC2. Mathematical Representations
  MC3. Clarity of Presentation

A. Problem Solving: The Essential Skill

The National Council of Teachers of Mathematics' *Agenda for Action* (1980) recommended that problem solving be the focus of school mathematics. Revisiting that recommendation, the *Curriculum and Evaluation Standards for School Mathematics* place becoming a mathematical problem solver at the top of the list of goals for students:

> *The development of each student's ability to solve problems is essential if he or she is to be a productive citizen... To develop such abilities, students need to work on problems that may take hours, days, even weeks to solve. Although some may be relatively simple exercises to be accomplished independently, others should involve small groups or an entire class working cooperatively. Some problems also should be open-ended with no right answer, and others need to be formulated.*

This goal is the foundation upon which Vermont has built its assessment program. Mathematics programs should be reducing emphases on the traditional one-or two-step problems that are categorized into traditional types, and moving toward a broader definition of problem solving. Problems should have a variety of structures. They should include the types of problems that students encounter every day. Application problems should play a major role in the curriculum. Some problems should be open-ended. Problem solving should include investigations and long-term projects. Teaching problem-solving strategies must be an integral part of instruction, and must be reflected in the process.

The NCTM Standards encourage a variety of problem-solving opportunities. The problem-solving assessment standard states:

> *"If problem solving is to be the focus of school mathematics, it must also be the focus of assessment. Students' ability to solve problems develops over time as a result of extended instruction, opportunities to solve many kinds of problems, and encounters with real world situations...*
>
> *Assessments should determine students' ability to perform all aspects of problem solving. Evidence about their ability to ask questions, use given information, and make conjectures is essential to determine if they can formulate problems.*

> *Assessments also should yield evidence on students' use of strategies and problem-solving techniques and on their ability to verify and interpret results. Finally, because the power of mathematics is derived, in part, from its generalizability (e.g., a two-space solution can be generalized to a three-space solution), this aspect of problem solving should be assessed as well."*

Vermont's commitment to providing meaningful problem-solving activities for its students was the basis for development of the problem-solving criteria.

Vermont's Problem-Solving Criteria

Too often problem solving has been taught as a linear process with four distinct steps: begin by restating the problem, identify a strategy, solve the problem, check your answer. This mechanistic approach survived for years as the way to teach students how to approach word problems. Vermont's concept of problem solving extends far beyond the simplistic approach to solving word problems, and is meant to assist students in developing meaningful approaches for the range of types of problems they will encounter in their lives.

Vermont educators also recognize that problem solving is not necessarily a linear process. Problems can have multiple solutions or multiple ways of obtaining a solution. Recognizing that students have different knowledge bases and varied learning styles suggests

> *Vermont's concept of problem solving extends far beyond the simplistic approach to solving word problems.*

that it is inappropriate to adopt a singular approach to problem solving and endorse that as the only approach valued by the state. The development of a range of problem-solving strategies (e.g, trial and error, listing, application of algorithms, visual representations) and a repertoire of problem-solving skills (e.g., reflection, analysis, verification) are the goals of mathematics education in Vermont. The problem-solving criteria adopted by Vermont reflect these goals.

Elements of problem solving are heavily integrated, and it is difficult to separate out distinct aspects. Nevertheless, Vermont's assessment must provide meaningful feedback to programs. To meet that goal Vermont isolated the following four key criteria for the problem-solving abilities of students:

Problem-Solving Criteria
• **Understanding of the task.**
• **How the student approached the task; the approach(es), procedure(s) and/or strategies adopted to attack the task.**
• **Why the student made the choices along the way; the reflection, justification, analysis, rationale, verification that influenced decisions.**
• **What findings, conclusions, observations, connections, generalizations the student reached.**

Although these criteria may suggest a sequence of activities, that is not necessarily the case. Portfolio entries have ranged from very simple problem-solving tasks to very complex multiple-week investigations. More complex problems offer opportunities to reach conclusions at various points through the problem-solving process. Similarly, a student may not begin with a full understanding of the problem, but through application of a strategy or futile attempts at verification, he/she may come to a fuller understanding of the task. The criteria serve as categories for classification of evidence — evidence that can be found throughout a student's solution to a piece, not necessarily at a particular place within the response, not as "parts" of a solution. A brief explanation of each of the criteria follows.

Understanding of the Task

It may go without saying that a student needs to understand what is being asked of him or her. In order to solve a problem you must understand the task. Understanding can include appreciating relevant information, being able to interpret the problem, and asking key questions that push for clarification. This criterion is a measure of the *receptive* communication skills of the student.

The rating scale for Understanding of the Task is:

1. Totally misunderstood
2. Partially understood
3. Understood
4. Generalized, applied, or extended.

At the lowest level, the student is not able to understand what is being asked. A blank response or a response that is not responsive to the task, or a clear misinterpretation of the task, are indicative of total misunderstanding. At the next level, the student might understand part of what is being asked or respond to one portion of the problem, while missing other key sections or critical information.

A level 3 response suggests that the student understood the task. Comprehension might be exhibited through a detailed description or analysis of the problem, or simply with a complete and correct response that reflects an understanding of the problem. The highest level of this scale suggests that the student stopped and analyzed the problem statement at the outset, and looked for special cases, missing information, or particular concerns, assumptions, etc., that might influence the approach to the problem.

It is important to note that understanding the task does not require a restatement of the problem. In fact, restating the problem in one's own words may not provide any evidence of understanding of the task. The understanding can emerge throughout the student's solution to the problem: Explanation of the task, the reasonableness of the approach, and the correctness of the response all provide evidence of a student's understanding of the task.

Quality of Approaches/Procedures/Strategies

Most problems have multiple ways in which they can be solved. Over time, students develop a repertoire of approaches, procedures, or strategies to solve problems. Strategies can include simple guessing, guessing and checking, systematic listing, using some form of manipulative, using Venn diagrams, using grids to record possible combinations, using formulas, and applying algorithms. There should not be just one way to solve a problem. Math teachers now recognize that it is more important to teach students

multiple approaches to problem solving, and to let them choose methods that work for them.

Although many different ways should be valued, selection of a strategy that can lead to an answer remains a goal. Students who select "guess and check" as a strategy for a problem that will take years to solve in that way should be able to evaluate the strategy and recognize that it is not viable, and they should select another approach.

The rating scale for Quality of Approaches is:

1. Inappropriate or unworkable approach or procedure
2. Appropriate approach/procedure some of the time
3. Workable approach
4. Efficient or sophisticated approach or procedure.

At the first level, the student has chosen an approach or procedure that will not lead to a solution for the task. The second level allows for the complexity of some tasks which will call on students to complete multiple tasks within the exercise. In the event that the approach or procedure is workable for some of the task, but not all, then the response is a level 2.

If the approach or procedure is viable and can lead to a solution, the piece is rated at a level 3. There are many routes to a solution, and each of these is treated as an equally acceptable strategy for this criterion. The most common approaches, as well as other,

---

We believe we must work toward increasing the attention students give to the process, as opposed to "the answer."

---

seemingly more cumbersome or inefficient responses, earn a rating of 3. There will be times when students provide very sophisticated strategies to solve a problem.

When rating a piece for the approach or procedure we tend to look at the demonstration, the description of approach, and the actual student products of drafts, scratch paper, and other artifacts of the problem-solving process. Problem solving cannot be "right answer" focused; the key to effective problem solving lies in the strategies one uses to attack the problem and the skills one uses to reflect on the process, to check one's work, and to verify one's decisions. In order to communicate the importance of the process, this criterion, which we sometimes refer to as the "how" of problem solving, emphasizes the approach and the viability of the strategies adopted by the student.

We recognize that students do not always record the procedures they follow, and we believe we must work toward increasing the attention students give to *the process*, as opposed to "the answer." We need to ask students "what are you doing?" and have them describe the process in precise terms. The importance of process in problem solving suggests that process is the answer to many of these tasks. It is also important to note that students do not always label their strategies (nor should they, necessarily), and raters must try to follow and label the student's approach based on the record of work that students keep. It falls to professional judgment to infer what the student adopted as an approach.

Why the Student Made the Choices Along the Way

Problem solving is more than understanding the task and selecting a viable strategy. Good problem solvers are constantly checking their assumptions, reflecting on decisions, analyzing the effectiveness of strategies, checking for exceptions, and verifying results in other ways. These skills provide an overlay for the problem-solving process. They may be the most difficult to teach, and they are clearly the most difficult to record — but they are critical to good problem solving.

These decision-making skills occur throughout a piece. They can affect selection of strategy, rejection of strategy, selection of focus, reflection on whether or not progress is being made, verification of steps, consideration of options, and other decision making. Whenever a student makes a conscious decision based on analysis, reflection, or verification, he/she is justifying the decisions or choices made along the way in the process.

Finding the *why* that underlies decisions is very difficult. Students rarely document and explicate their decision making. Statements in portfolios like "I realized this process would not work" or "I knew there was a quicker way," or "This has to be a solid approach because," or "I thought I was right but realized I was wrong because..." all provide evidence of the metacognitive skills associated with problem solving.

The rating scale for why the students made the choices along the way reflects the difficulty in capturing the evidence for this criterion. Students are getting better at explaining **how** they solve a problem, the approach they followed to solve the problem, and the steps along the way. Getting them to think about why they proceeded they way they did, and to communicate this process orally or in writing, is a bigger challenge. The rating scale for this criterion is:

1. No evidence of reasoned decision making
2. Reasoned decision making possible
3. Reasoned decision making/adjustments inferred with certainty
4. Reasoned decision making/adjustments shown/explicated.

The scale is dependent on the rater's ability to combine professional judgment with inference. The extremes of this scale are evident. At the bottom is the student who attempts the problem without ever making informed decisions. He/She simply does the problem, attacking it in a seemingly random fashion and never reflecting on options or decisions, evaluating alternatives, verifying decisions, or thinking about the selections made throughout the process. Absence of any evidence of informed decision making leads to a level 1 rating. At the other end of the scale, a student clearly articulates the decisions made, either through explanation or example.

The issue is not what did you find, it's *so what* does that mean?

In between, it is often difficult to determine whether or not the student has engaged in reasoned decision making. A student seems to begin with one approach and then switches to another approach. Did he/she make the change because he/she recognized the first approach wouldn't work, or did he/she tire in one approach and decide to try another? A student seems to reach an acceptable response, but then begins to solve the problem another way. Is he/she verifying the answer, or does he/she believe the first approach was wrong?

In each of these examples it is possible that the student is thinking about the process, reflecting on his/her problem solving, and making adjustments when necessary. It is also possible that he/she is not! The student's actual product often provides guidance to a rater as to which of the two options is most likely correct. If the product shows a pattern that teachers have seen through years of

teaching — for example, begin with random listing and then switch to a more systematic guess and check — it is likely that reasoned decision making occurred. However, an attempt to infer decision making suggests that although reasoned decision making may have occurred, it is equally likely that it didn't. When you can infer decision making with some level of certainty the response is a level 3; if it may have occurred but is equally likely that it didn't, then the response is a level 2.

The third criterion is clearly the most difficult to capture. It may also be the most important component of problem solving. We are not accustomed to stopping the creative process of problem solving to examine what kinds of decisions we are making, and why we make them. It is uncommon, too, for people to document the process. However, it is important that students be able to analyze their own decision-making skills and, particularly in group problem-solving situations, to be able to share the decisions and justifications with other group members, so other students can not only follow the process but also critique, evaluate the reasonableness of the decision, suggest alternatives, and make other metacognitive analyses. The challenge associated with this criterion is how to capture the process without interrupting the problem solving. It is important to note that students do not have to document the decision-making process in every piece of work, but they should be able to do so, and the documentation should be reflected in their best pieces.

What Decisions, Findings, Conclusions, Observations, Connections, and Generalizations the Student Reached

A goal of problem solving is to reach a solution, but getting an answer is less important than making connections or extending the solution. Mathematics is no longer about finding the answer to an exercise that is an artificial problem existing primarily for the purpose of testing problem-solving skills. The issue is not what did you find; it's *so what* does that mean?

Tasks should provide students with an opportunity to extend beyond their solution. Students should be encouraged to make observations about their conclusions, or to make connections to other mathematical concepts, to real world applications, or to other disciplines. This criterion requires students to ask *so what* at the completion of each problem.

The rating scale for the criterion is:

1. Solution without extensions
2. Solution with observation
3. Solution with connections or application
4. Solution with synthesis, generalization, or abstraction.

A level 1 response requires a solution. Correctness of the solution is not an issue. As students improve in their performance on the first three criteria, the likelihood of incorrect responses will be diminished. The bottom of this scale suggests that the student the problem and stops. Any attempt to question what the solution means, or to make an observation about the solution, leads to a rating of 2.

If the student goes beyond a simple observation and makes connections to other mathematics, to other disciplines, or to other possible applications, then the rating is a 3. In some instances a task will provide an opportunity for a student to synthesize information, or to come to some generalization or level of abstraction based on the observations made throughout the problem. In these instances the work is a level 4.

P6

Sums	Tally	
2	/	\
3		O
4	//	2
5	LHT LHT	10
6	LHT/	5
7	LHT LHT //	12
8	LHT /	6
9	LHT //	7
10	\ //	3
11	LHT /	5
12		O

What happened? The sum I got
the most is seven. I didn't get
any sums of three and twelve

Why do you think you got these
results?

P7

Sums	Tally	
2		O
3	/	1
4	LHT	5
5	LHT LHT	10
6	/	1
7	LHT LHT /	11
8	LHT /	6
9	///	3
10	//	2
11	//	
12	//	2

What happened? The sum I got the most
is number seven. I didn't get any sums
of two. I got sevens more than the others
ones because you can get six ways on
dice to get seven. I got two the least
because there is only one way to get
two the way to get two is one and one.

B. Communication: A Critical Element

The 1989 *Standards* of the National Council of Teachers of Mathematics included "Learning to Communicate Mathematically" as one of its new goals for students.

NCTM suggested that a student should learn "the signs, symbols, and terms of mathematics. This is best accomplished in problem situations in which students have an opportunity to read, write, and discuss ideas in which the use of the language of mathematics becomes natural. As students communicate their ideas, they learn to clarify, refine, and consolidate their thinking."

Vermont teachers support NCTM's belief in the importance of good communication skills, particularly in mathematics. It's only natural to include communication as a critical element in the assessment of student work and its evaluation in the context of problem situations. Recognizing that students will work with others in school and in life makes it essential that they learn more than independent problem-solving skills. They must be able to communicate their ideas to others, and understand the ideas of others, in order to fully benefit from the value of group analysis and reflection, which can further strengthen their individual problem-solving skills.

Students need to learn to communicate complex ideas. The emphasis on getting students to articulate how they solved the problem and why they made the choices they made throughout the problem call for sophisticated communication skills. Being an effective problem solver without being able to communicate ideas

to others limits mathematical power in a way that is inconsistent with Vermont's goals. For these reasons, communication is the second key element that will be examined through the Vermont portfolios. Again, note that the different types of problems provide opportunities for the use of different communication skills.

The NCTM assessment standard for communication is a bit more prescriptive than the problem-solving standard. It states:

...Integral to this social process is communication. Ideas are discussed, discoveries shared, conjectures confirmed, and knowledge acquired through talking, writing, speaking, listening, and reading. The very act of communication clarifies thinking and forces students to engage in doing mathematics. As such, communication is essential to learning and knowing mathematics. But communicating mathematically presents unique difficulties for students. Mathematics is heavily based on the use of symbols and attaches specific, and sometimes different, meanings to common words.

An assessment of students' ability to communicate mathematically should be directed at both the meanings they attach to the concepts and procedures of mathematics and their fluency in talking about, understanding, and evaluating ideas expressed in mathematics. As in any language, communication in mathematics means that one is able to use its vocabulary, notation and structure to express and understand ideas and relationships. In this sense, communicating mathematics is integral to knowing and doing mathematics.

Vermont's standard goes beyond the surface level of mathematical signs, symbols, and vocabulary. Communication must include the language of mathematics and mathematical representations (e.g., graphs, tables, charts, models, manipulatives), which provide a vehicle for presentation of ideas. General principles of communication (e.g., clarity, organization, provision of detail) must also be reflected in the rating of student work. These beliefs led to the identification of three criteria that comprise the communication strand.

Communication Criteria

Historically, mathematics education has focused on whether or not students could reach the right answer. How well they communicated their ideas, thoughts, and processes seemed secondary. Mathematics terminology, signs, symbols, and mathematical representations, including charts, tables, graphs, manipulatives, and diagrams, have been regarded by many as an end of mathematics (i.e., can a student graph and read data presented in a chart) rather than as a way of communicating mathematically. The way in which a student presents his/her work is yet another way in which he/she communicates ideas to peers and to the teacher. Oral and written communication are not solely the responsibility of the language arts teacher; mathematics teachers must also help students become better speakers and better writers. These understandings and beliefs led to the following communication criteria.

Communication Criteria
- Language of Mathematics
- Mathematical Representations
- Clarity of Presentation

Unlike the problem-solving criteria, which were rated on scales that measured the qualitative nature of the response, the rating scales for these criteria are more quantitative. The first step in language acquisition, mathematical or other, is trying to use the words. As a result, each of these scales begins with "no" use of these skills, and builds through "some" use to "sophisticated" use. The variety of tasks offered in programs creates a range of opportunities for students to demonstrate their skills, but the absence of opportunity to use these skills also makes a statement about the program. Programs must value and teach communication skills. A brief explanation of each of the criteria is provided in the section that follows.

Language of Mathematics

Vermont recognizes that mathematics is a language, and that students gain power in their ability to communicate with one another and with the community of mathematicians as they grow in their facility with the language. A student who never adopts the terminology will find it cumbersome to explain his/her approach in the absence of mathematical terminology or symbols. A student who talks about "timesing two numbers" needs to adopt correct terminology and speak about either multiplication or products. A student who can write a series using proper notation rather than speaking about the sum of 12 numbers that are the squares of the numbers from 1 to 12 inclusive is adept at using the language. Vermont's goal is to help students become more proficient in the use of mathematical vocabulary, notation, symbols, and structure to represent ideas and to describe relationships.

The scale for this criterion includes:

1. No or inappropriate use of mathematical language
2. Appropriate use of mathematical language some of the time
3. Appropriate use of mathematical language most of the time
4. Use of rich, precise, elegant, appropriate mathematical language.

A level 1 response suggests that the student rarely or never uses mathematical terms or notation to express ideas as part of the tasks. At the second level, students demonstrate some knowledge of the language of mathematics. They are beginning to use the language and notation sporadically, but often use words or non-mathematical terms in instances in which a mathematics term, symbol or notation would have been more precise and more concise.

> *Being an effective problem solver without being able to communicate ideas to others limits mathematical power.*

As students become better mathematical communicators, the language of math becomes a native language. When most of the communication is in the native language, the student is at the third level with respect to this criterion. Finally, some students find ways to use the language and its notation almost to the point of eliminating the need for other words. The use of symbols and notations in conjunction with representations and transitional phrases marks the rich, precise use of mathematical language that is the highest rating for this criterion.

Mathematical Representations

Beyond the terminology and notation, communication power in mathematics comes from being able to communicate information through graphs, charts, tables, diagrams, models, or some type of manipulative. Visual presentation in a format that has been accepted in the profession often communicates information more powerfully and more concisely than pages of text. Students need to be able to read information presented in diagrams, charts, bar graphs, line graphs, or pie graphs or through models. They also should be able to construct these representations and make connections that link information presented in different representations, including equations or formulae.

The rating scale for this criterion includes:

1. No use of mathematical representations
2. Use of mathematical representations
3. Accurate and appropriate use of mathematical representations
4. Perceptive use of mathematical representations.

The structure of this criterion is very similar to the mathematical language criterion. It represents a hierarchy of skills. At the first level, students make no attempt to integrate mathematical representations into their problems. The next level suggests that students are beginning to work with representations, and that charts, tables, models, and diagrams are beginning to appear throughout their work. At this level the key distinction is that the representations are used; accuracy and appropriateness are not factors. The accurate and appropriate use of representations is what distinguishes a level 3 from level 2 response. Sophisticated use of representations at level 4 is evident when the student begins to recognize the power of representations, using the best methods for communicating ideas and presenting them in simple but elegant mathematical ways.

Clarity of Presentation

The final criterion for communication addresses the general communication skills of the student. Although this criterion is not mathematics in the strictest sense, the importance of oral and written communication skills demands that they be assessed across the curriculum. Students working in groups with other students or trying to communicate their ideas to teachers and parents must be able to organize their thoughts and present them to others in a format that is organized, coherent, and sufficiently detailed to allow another individual to follow the student's thinking. The presentation of ideas is at the core of this criterion, and it is an essential element of portfolio-based tasks.

The scale for this criterion includes:

1. Unclear (disorganized, incomplete, lacking detail)
2. Some clear parts
3. Mostly clear
4. Clear (well organized, complete, detailed).

The primary indicator of where a response rates on this scale is the amount of inference required on the reviewer's part. If you cannot follow the student's thinking or processes at all, or if insufficient work is provided to give you the level of detail needed, then the response is at level 1. The fourth level indicates that the student has structured the response well and provided detail throughout the presentation. The rater did not need to fill in missing information or make inferences about how the process evolved.

The primary distinction between the middle two levels is related to the amount of inference required from the rater. If the student has presented most of the information in a clear way, only occasionally relying on the rater to fill in steps, then the rating is a level 3. On the other hand, if there are some clear parts, but the rater needs to fill in the missing steps in several sections, the rating is a 2.

The communication criteria are similar to the problem-solving criteria in that they are all interrelated — and it is critical that teachers design tasks that provide an opportunity to learn these skills as well as demonstrate them. At times it will appear that these criteria overlap with the problem-solving criteria. They do. The communication criteria provide an opportunity to evaluate the tasks from a different perspective.

Applying the Communication Criteria: Some Examples

The quality of communication within student work is closely related to the problem-solving criteria. The better a student communicates, including use of mathematical terminology, representations, and the clarity of presentation, the greater the level of certainty with which teachers can evaluate the problem-solving ability. Read through C-1's "Mr. Fister" problem and focus on his use of mathematical language and the clarity of his thinking. Notice he has provided a rough draft and a final product.

This student uses mathematical language appropriately most of

the time. His use of terms like fractions, common denominator and numerators provides evidence of his skills. In a few instances he could be more precise, like using "subtract" instead of "take away." He does not use math representations. In terms of clarity, the piece is well organized, complete and detailed.

Compare C-1's approach to C-2's. Notice that C-2's approach is exclusively algebraic, with no words, no explanation, no illustrations. It has some clear parts, but is not as rich in language or in clarity of presentation as C-1's.

One aspect of mathematical communication is the ability to represent a concept in many ways. C-3's illustration of ways to show a number just begins to convey this power. Mathematical representation focuses on the use of models, diagrams, manipulatives, graphs, tables and charts to communicate information or concepts. C-4 used a diagram to show how 4 three cent purchases of peanuts produces 28 peanuts but 6 two cent purchases (also 12 cents) produces 30. The representation shows C-4's thinking and communicates why the 5 for two cents is a better deal. It is accurate and appropriate use of representations.

C-5 also tried to use mathematical representations for the peanut problem, but the chart doesn't relate to his answer. This response shows use of mathematical representations, but it is not accurate and appropriate.

Sometimes, mathematical representations stand alone. C-6 solved a problem regarding crossing a stream through diagrams alone. This response made perceptive use of mathematical representation.

C-7's response to the survey problem shows the use of charts, bar graphs, and line graphs to communicate information. This perceptive use of mathematical representation shows the level of performance we would like to see from students.

The clarity of the portfolio entries addresses the general communication skills of the student. Presentations should be well organized, complete, and detailed. Both examples C-8 and C-9 demonstrate how clearly students are able to organize their thoughts, incorporating mathematical language and representations when appropriate.

Composite Ratings

The rating scales have been defined and illustrated in terms of individual portfolio entries. The same scales are used for final composite ratings of the best pieces designated by a student as his/her best piece. Raters review the best pieces, consider the overall quality of the student's work, and provide a composite rating for each criterion. These ratings are the basis for state reports. The dual purpose of scales (i.e., for individual pieces and for collections of best pieces) places some limitations on their use.

The more narrow an entry, the more difficult it may be to score. Good portfolio pieces are open-ended, may be ongoing, and provide students with multiple opportunities for response. The richer the task, the better the "fit" to the criteria. However the combination of from a number of pieces builds the reliability of the composite, or portfolio rating for best pieces.

It is also important to note that although the same criteria and same scales are used at each grade level, the interpretations of the scales differ at each of the grades. The expectations for a Level 3 performance for math representations for grade 4 are not the same for grade 8. The decision rules which guide rating and the models that illustrate the markers vary by grade level.

III. Portfolios as a "Window" on Math Programs

Portfolios of student work provide a way of looking at mathematics programs and assessing the *instructional opportunities*, the *content areas*, and the *level of empowerment* of students reflected within the programs. This section provides a description of these three elements.

A. Instructional Opportunities

The following five instructional opportunities are critical aspects of effective mathematics programs:

- Group work,
- Interdisciplinary work,
- Manipulative use,
- Real world application, and
- Technology use.

NCTM standards call for increased opportunities for individual and group assignment, and for discussion between students and teachers and *among students*. Students must learn to work cooperatively in groups, to learn from one another, and to recognize the power in building upon other's ideas, critiquing and receiving criticism from others. Mathematics, and problem solving in particular, offers a vehicle for building skills as members of groups. Portfolios should include individual and group work.

Integrating mathematics with other disciplines provides opportunities for students to explore mathematics from different points of view. When they make connections to art, science, music, business, industrial arts, social sciences, and other disciplines, students' appreciation of math can be heightened. Interdisciplinary work should be reflected in portfolios.

Children construct mathematical understandings by manipulating concrete objects. Although this is particularly true for elementary school programs, it is also important at the junior and high school levels. Cubes, base ten and pattern blocks, tiles, geoboards, balances, graph and grid paper, and counters like buttons and beans should be a part of math programs, and evidence of work with them should be reflected in some way in portfolios.

The real-world emphasis of mathematics must be reflected within the programs. Mathematics cannot be taught as an abstract field, with artificial problems designed to test a student's ability to learn concepts. Problems should be meaningful, with multiple solutions, and should be complex enough to challenge students. The "real-world" nature of the problems and the generalizability of skills to other contexts should be evident throughout the program, as evidenced in portfolios of student work.

Finally, technology must be a part of the mathematics program. Students at all levels should have access to calculators and computers. Technology should be integrated into the curriculum. Students should also have the freedom to bring technology into their approaches to problem-solving tasks. Their assignments and their products are a source of evidence of how well technology has been included within a program.

A review of a representative sample of portfolios from a program should provide evidence of each of these five instructional opportunities.

B. Content Areas

Mathematics programs must include a variety of content as part of the curriculum. NCTM specified a number of content standards in addition to problem solving, communication, reasoning, and connection standards at each grade level. The Vermont Portfolio Committee specified content expectations for each of the two grade levels included in the portfolio assessment. Programs at the fourth and eighth grade should provide students with opportunities to address problems that require the skills and content knowledge associated with each of these areas. The specific content areas are:

Grade 4	Grade 8
Number Sense — Whole No./	Number Relationships/
Fractions	Number Theory
Operations/Place Value	Estimation
Estimation	Patterns/Functions
Patterns/Relationships	Algebra
Geometry/Spatial Sense	Geometry/Spatial Sense
Measurement	Measurement
Statistics/Probability	Statistics/Probability

Categorization of portfolio pieces from a sample of complete portfolios for a classroom into these eight distinct content areas provides an indication of the content range of programs within the state.

C. Disposition/Empowerment

Mathematics in the 90's must be for all students, not those few who aspire to math-related careers. All students must learn to value math and become confident in themselves and their ability to be mathematicians. Everyone is a mathematician. Estimating the total bill at the grocery counter, knowing how much paint to buy to cover the deck, wrapping a birthday gift, monitoring fuel efficiency, considering the approprate tire size, and following a recipe all require skills of mathematicians. Math programs must be designed to help students succeed. The classroom climate must foster and demand risk taking. Perseverance should be rewarded. Flexibility should be valued. Reflection must be required. As mathematical programs change, teachers must attend to student disposition towards mathematics, not just cognitive understanding.

It is with these goals in mind that Vermont added the third element, Mathematical Empowerment, to Problem Solving and Communication as the three main elements for reviewing individual student work. This element is also responsive to NCTM's recognition of disposition as a key standard for assessment. The NCTM Assessment Standard reads:

Learning mathematics extends beyond learning concepts, procedures, and their applications. It also includes developing a disposition toward mathematics and seeing mathematics as a powerful way for looking at situations. Disposition refers not simply to attitudes but to a tendency to think and to act in positive ways. Students' mathematical dispositions are manifested in the way they approach tasks — whether with confidence, willingness to explore alternatives, perseverance, and interest — and in their tendency to reflect on their own thinking.

The concept of empowerment recognizes that mathematics, and math teachers, must look beyond the cognitive aspects of programs and attend also to affective issues. The development of mathematical power in students will require that teachers maintain a classroom environment conducive to problem solving, and that teachers model the curiosity, flexibility, and reflection that we must instill in students.

Good problem solving requires a disposition towards curiosity. Students must wonder about options, think about what might happen if they choose one path instead of another, and be curious enough to want to explore each of the alternatives and find out what happens. They must be flexible in their approaches, recognizing that not all paths will lead to solutions, but each may lead to a distinct adventure. Students must be risk takers, not always following the "safe" path that they've tried before. They should be ready to try new ideas and new alternatives. They need to know that this behavior will be rewarded by the teacher, not viewed as bad decision making or ineffective use of time.

Problems should be meaningful, with multiple solutions, and should be complex enough to challenge students.

Students also must learn to persevere on problems, not walk away at the first sign of frustration. Individually, and as a part of a group, they should learn how to persevere, to see problems through to conclusions. As part of the process, students must also learn to step back from the process and reflect on their thinking, analyze the steps they have taken and think about how to proceed. They must learn along with their peers and their teachers how to be reflective learners.

Developing these dispositions (curiosity, flexibility, risk-taking, perseverance, and reflection) in students will help them develop as mathematicians. The dispositions will lead to greater success as problem solvers, which in turn build their confidence and increase motivation to grow in mathematics. This success will help them grow in the ways in which they value mathematics.

This third element of individual portfolio entry ratings does not lend itself to rating scales in the ways that problem solving and communication do. It is best viewed in the classrooms and in the teaching of mathematics. However, unless we begin to ask students questions that get them to start reflecting on ways they think about mathematics, they won't begin to see the changes within themselves with respect to mathematical empowerment. Portfolios should provide an opportunity for students to write about ways in which they are growing in mathematics, and how they feel about the discipline.

Evidence of empowerment is often found embedded in the narrative that accompanies individual problem-solving entries.

Some teachers have asked students to comment on how they felt after completing the problem. Statements like "This seemed to take forever, and at times I thought I'd never solve it, but when I finally realized...it felt great," "In order to figure this out first I drew a picture, because learning with pictures is easier for me than numbers..." "I thought I had a right answer, but I started to wonder what would happen if I..." and "I was sure I finally did it because..." provides some evidence of the perseverance, curiosity, and confidence that students begin to feel and then communicate. Writing in response to questions like "What color is math?" or "Where am I ever going to use math anyway?" provide additional information. Some teachers require students to keep a mathematics journal. Entries within the journal often provide insight into the affective growth of students in mathematics.

Empowerment is not "scored"; anecdotal information will be collected to summarize programs across the state. A review of these data may lead to systematic rating in the future.

A few examples of student writing that illustrate the heart of this element are provided on the pages that follow.

Writing and Learning: Some Samples

As students begin to write about the work they do in mathematics, evidence begins to emerge about how students feel about their work, and how they value mathematics. Some students, like the author of example **E-1**, write reactions to individual pieces of work. This reaction to a short problem communicates his frustration and the student's reaction to a recent math test on which he did poorly.

"As you can see...bad mood, I just got a 45 on my math test."

Over time, E-1's perception about problems began to change. A comment, which he labeled "Opinion Corner" on a later problem, suggests a problem was "kinda fun."

Everyone is a mathematician...Math programs must be designed to help students succeed.

"...I would recommend this problem for any one who enjoys math and thinking."

Other students write in mathematics journals. The samples were taken from a fourth grade class. One journal includes responses to questions asked by the teacher. Other journals provide an unstructured approach to elicit student writing about mathematics.

After completing an application problem in which she forgot a critical factor, another student made the following observation: "Even though using the correct numbers will get you an answer it may not be the correct one! Calculators don't know that we are making book covers and don't want to glue together a whole bunch of pieces!"

VI. The Vermont Mathematics Portfolio Program: What We Have Learned and What We Need To Do

The pilot year was very instructive. The process of reviewing portfolios from more than 40 schools led to several observations about the program, and served as the basis for decisions regarding program implementation. Here are the observations:

1. Mathematics portfolios do provide a window on mathematics programs.
    Mathematics portfolios work. Although the portfolios reviewed represented a broad range in terms of quality of entries, it is clear that the potential exists to develop a program in which students put forth their best efforts at problem solving, in a format that is not a standardized test. Many teachers experimented with new types of instruction and assessment, and documented student work in portfolios that were rich in content. The promise of the program is evident, and heartening rewards were reported by teachers who wrote about growth in their students, of themselves as teachers, and of their mathematics programs. The challenge is to provide adequate direction and support for other teachers as they work to adopt portfolios as part of their assessment.

2. The problem-solving and communication criteria and scales work.
    Committee members report that the scales work. The criteria worked well with problem-solving pieces that are appropriate portfolio entries. *They did not work, and should not work, with computation sheets, drill sheets, and other student products that would not be categorized as problem solving.* Committee members found few examples of problem solving that did not lend themselves to scoring with the criteria. The language of the scales has intuitive appeal to most teachers. Teachers report that the language of the scales denotes distinct levels of performance, which makes the scales easy to adopt and use with very limited training.
    The challenge for future years is to get mathematics programs to change their emphases, as recommended in the NCTM Standards, away from drill and practice and computation and toward more open-ended problem-solving tasks. This is the type of instructional change that Vermont is seeking. Evidence of this change will come from portfolios of student work.
    Some refinement of the application of the scales will be necessary as teachers strive to meet Vermont's instructional goals and assessment goals. A general rule for developers of standardized tests is that testing should not be designed to teach. The rule breaks down when you try to capture artifacts of everyday instruction as a basis for testing. Teaching students to learn to reflect on decisions — to ask questions like "How will this help me?", "Is this moving the process forward?", "Did I approach that reasonably?", and "How else could I check this?" — requires that teachers model the questions. Over time, students learn to assume more responsibility for the process. The scoring system must accommodate these varying levels of independent performance with respect to the criteria. The state must clarify scoring rules to recognize the need for structure, for example rating prompted performance slightly lower than unprompted performance. Teachers need to know they need to provide reflective structure; but over the course of the year, students should begin to internalize

the prompts.
    The scales appear to work at the general level. As the program develops we will need to establish decision rules specifying how to apply the scales to special types of problems, like the one explained in the previous paragraph. There will be other special cases. If a fourth grade teacher provides a very structured problem, how do you rate clarity? How are omissions or missed opportunities reflected in the scoring? If a teacher asks for a structured observation, is the student credited for reaching a generalization? Standard responses to these and other questions must be generated to ensure consistent scoring of portfolios. This refinement and clarification of how the scoring criteria are applied to specific cases must be a focus for the coming year.

3. Additional specification of portfolio content is necessary to provide equitable basis for evaluating student performance.
    Each teacher and each school took a different approach to assembling portfolios of student work. Some had single entries; some had dozens. Some portfolios were limited to two or three special projects developed for the portfolio; some included every homework assignment, worksheet, and test the student had taken. A few portfolios were constructed exclusively from puzzles and challenge sheets; others were limited to investigations. As committee members reviewed the various approaches to assembling portfolios, it became apparent that the state needs to be more prescriptive in terms of the types and amount of content that should be included to be fair to programs. The committee made the following recommendations regarding best pieces:

- A minimum of five and a maximum of seven best pieces should be identified within the portfolio.
- Among the five to seven pieces, at least one must be a puzzle or non-routine problem, at least one must be an investigation, and at least one must be an application.
- The best pieces should include a sample of student writing in math (e.g., essay, journal, report).
- Entries should include pieces that allow for a range of performance, not all simple activities that merely require students to apply skills they already know.
- Group work should be reflected among the best pieces, but no more than two entries may be exclusively group products.
- Best pieces must be limited to tasks that are indicative of problem-solving activities. Computation sheets, drill sheets, and homework assignments that do not meet these criteria are not acceptable as best pieces. Teachers should assure that the piece is scorable with respect to the criteria.
- Selection of best pieces should be completed by the student in consultation with the teacher. An introduction to the portfolio, including a table of contents and an explanation of why the pieces were selected, should be prepared by the student.

    These specifications will help to even the playing field for students as they try to demonstrate their knowledge and skills in a way that is consistent with the state's criteria and comparable to their counterparts across the state. The process still allows teachers to specify the content of their program, and to meet the state goals in many different ways.

The committee made the following recommendations regarding the whole portfolio. In addition to the pieces selected by students as their best work, the portfolio should include the range of work that comprises the program. Portfolios should include evidence of:

- Group work,
- Use of manipulatives,
- Use of technology,
- Connections to other disciplines, and
- Connections to real world application.

These specifications ensure that a complete picture of the program is provided, not just the nature of the work that was the students' best.

4. Portfolios need additional structure to facilitate scoring.

The scoring of portfolios requires some standardization in terms of structure and presentation. Teachers and students must be trained in the process. Students must understand the purpose of portfolios, and take ownership of them as products of their work. The goals of mathematics programs, as articulated through the criteria and rating standards, must be clear to students so they have a sense of the purpose of their instruction and the state's assessment.

Students must work with their teachers to review their portfolios on a regular basis, organizing the entries and selecting the 5-7 best pieces of their work with respect to the state's criteria. They must understand that the pieces they select must stand alone. A copy of the task must be included, and the student's approach to the problem must be provided with a clear and complete description of the way he/she completed the task. A table of contents for the entire portfolio and a letter from the student that lists the key pieces and focuses on why they were selected will provide the context necessary for state raters to evaluate the work.

5. Mathematical programs must be held to high standards.

The quality of student work depends upon the opportunities provided by the teacher as part of the instructional program. Students in classes where teachers provide problem-solving tasks that lend themselves to connections and generalizations will have the opportunity to score higher than students in classes that are not given that opportunity. This is not an inequity in the scoring system. If an instructional program does not provide an opportunity for students to learn to make connections or extensions and another program does, the inequity rests with the programs, not the rating system. The portfolio system provides a vehicle through which these inequities can be identified and, through the support system, corrected.

Similarly, the range of responses possible on a given piece of work may have a floor or a ceiling with respect to the scale for a particular criterion. If a teacher provides a piece that limits a student's response, there will not be an opportunity to score highly on several criteria for that piece. A more difficult task might provide opportunities to score highly on several problem-solving criteria, but may not include opportunities to demonstrate mathematical representation. Each of these possible criticisms speaks to the need for diversity within the portfolio, which can only be achieved if there is diversity within the mathematics program. Again, equity for students must come from comparable standards of quality for all mathematics programs.

6. Mathematics instruction must change.

The review of portfolios and the application of problem-solving

and communication criteria to the work confirmed our belief that instruction needs to change. These criteria reflect a new way of thinking about and of teaching mathematics. Problem solving must be the focus of programs. Problems need to be meaningful, with real world applications. Students need to become more adept at examining their own problem-solving strategies and communicating their thinking to others. They need to learn to ask "so what" when they think they have solved the problem. Mathematics programs must also begin to emphasize the language of mathematics and the use of mathematical representations. Clarity of thought and writing are as important in mathematics as they are in other content areas.

Programs must begin to work toward nurturing the dispositions necessary to make students confident in their knowledge of mathematics and to value mathematics. Group work, interdisciplinary work, use of manipulatives, real-world applications, and technology must be reflected in all math programs.

These expectations reflect the types of changes necessary to ensure that programs are responsive to changes in the teaching of mathematics. The mathematics portfolio assessment program is a new way of capturing information about student performance data, and an integral part of a revised approach to instruction. It will guide the types of changes necessary in instruction, and provide a vehicle for tracking the change.

7. Changes in instruction and assessment require additional resources and training.

In order to effect the types of changes required in instruction and assessment, schools, teachers, and students will require support, including professional development opportunities that engage teachers in reform and assist them in identifying and developing new strategies, resources, and materials that will help them as they re-evaluate their own programs and teaching. Professional development must address such diverse issues as problem-solving skills, teaching students to be better problem solvers, teaching writing across the curriculum, collaborative learning, integration of technology into the curriculum, creating a classroom environment that is conducive to problem solving, and working across disciplinary boundaries. In addition to professional development on instructional issues, teachers also need training in using portfolios, identifying tasks, integrating problem solving into the curriculum, and training in the scoring system.

The nature and level of change sought will require resources. The state will develop a regional network so teachers can work together to develop the instructional and assessment practices necessary to make this program work.

What will it take for this program to work?

During the summer of 1991, workshops are being offered throughout the state to assist schools and teachers interested in using portfolio assessment in their classrooms. If the state chooses to adopt portfolio assessment statewide, the achievement of its goals will depend on successful implementation of the following key ideas, which emanated from the pilot year:

1. Provide additional professional development for changes in math instruction.

The Mathematics Portfolio Criteria reflect changes underway in mathematics curriculum, evaluation, and teaching. The creation of a statewide assessment system will not automatically lead to the types of changes necessary in mathematics education. Teachers and schools need support to change their programs and their teaching to reflect the revised standards. This support must include professional development related to teaching problem solving, resources for modifying curriculum, ideas for interdisciplinary teaching, ways of integrating technology into the classroom, and new ways of evaluating students. This support can be achieved through statewide and regional workshops; ongoing support can come through regional networks of math resource consultants.

2. Create regional networks to provide support and training.

Although the state can provide support for the changes in instruction, real change will demand local support. Decentralized, regional networks will provide the localized support necessary to effect and institutionalize the changes. Teams of teachers trained and supported by the state program who are available to work with their colleagues on a regular basis in their own classrooms hold the greatest potential for program success.

3. Communicate portfolio criteria early in the year.

The criteria have undergone significant development during the pilot year. The program must begin to stabilize, and the criteria for problem solving and communication must be accepted as "final criteria" for the next several years. These criteria and the standards defined by each scale must be communicated to all mathematics teachers at the beginning of the school year. The focus in the early months of the school year should be toward developing a deeper understanding of the criteria and how they apply in a teacher's individual classroom. Summer institutes for teachers and regional meetings early in the fall will address this need.

4. Train all fourth/eighth grade teachers in the scoring of portfolios according to state standards.

Teachers must be able to score their students' work according to the state criteria. This suggests that two levels of training will be needed in the coming year. First, teachers must learn about the portfolio criteria, how to design instruction that will map to the criteria, and how to keep portfolios as part of their instructional programs (as described in goal #3). Second, their own standards must be calibrated to the state standard for each criterion. Every fourth and eighth grade mathematics teacher in Vermont should be able to review student work, from his/her own class or that of a colleague, and reliably rate the work according to the state standard. This will require extensive training and recalibration throughout the first year.

5. Provide resources to facilitate the integration of this assessment with instruction.

As noted in the first four goals, it is clear that two distinct programs are being instituted. New directions and new goals have been established for math education within Vermont, requiring instructional change. Portfolios are an extension of this change, and will also provide the state with a vehicle for evaluating programs. The instructional change, assessment change, and institution of state evaluation must all be integrated. Portfolio assessment must be integrated with instruction; teachers should not view portfolios as something they do in addition to instruction to meet a state need. Teacher handbooks, resource guides, ideas for managing portfolios, and guidance in the use of portfolios of student work as part of the instructional program are key to making the program work. Portfolios must be perceived by teachers as feasible and useful to their practice.

Appendix C
Mathematics Rating Form

	A1 Understanding of Task	A2 How – Quality of Approaches/Procedures	A3 Why – Decisions Along the Way
Student: _____ ID Number: _____ School: _____ Grade: _____ Date: _____ Rater: _____	**SOURCES OF EVIDENCE** • Explanation of task • Reasonableness of approach • Correctness of response leading to inference of understanding	**SOURCES OF EVIDENCE** • Demonstrations • Descriptions (oral or written) • Drafts, scratch work, etc.	**SOURCES OF EVIDENCE** • Changes in approach • Explanations (oral or written) • Validation of final solution • Demonstration
ENTRY 1 Title: _____ P I A O Puzzle Investigation Application Other			
ENTRY 2 Title: _____ P I A O Puzzle Investigation Application Other			
ENTRY 3 Title: _____ P I A O Puzzle Investigation Application Other			
ENTRY 4 Title: _____ P I A O Puzzle Investigation Application Other			
ENTRY 5 Title: _____ P I A O Puzzle Investigation Application Other			
ENTRY 6 Title: _____ P I A O Puzzle Investigation Application Other			
ENTRY 7 Title: _____ P I A O Puzzle Investigation Application Other			
OVERALL RATINGS →	**UNDERSTANDING OF TASK** FINAL RATING 1 Totally misunderstood 2 Partially understood 3 Understood 4 Generalized, applied, extended	**HOW – QUALITY OF APPROACHES/ PROCEDURES** FINAL RATING 1 Inappropriate or unworkable approach/procedure 2 Appropriate approach/procedure some of the time 3 Workable approach/procedure 4 Efficient or sophisticated approach/ procedure	**WHY – DECISIONS ALONG THE WAY** FINAL RATING 1 No evidence of reasoned decision-making 2 Reasoned decision-making possible 3 Reasoned decisions/adjustments inferred with certainty 4 Reasoned decisions/adjustments shown/explicated

COMMENTS:

A4 What – Outcomes of Activities	B1 Language of Mathematics	B2 Mathematical Representations	B3 Clarity of Presentation	CONTENT TALLIES
SOURCES OF EVIDENCE • Solutions • Extensions – observations, connections, applications, syntheses, generalizations, abstractions	SOURCES OF EVIDENCE • Terminology • Notation/symbols	SOURCES OF EVIDENCE • Graphs, tables, charts • Models • Diagrams • Manipulatives	SOURCES OF EVIDENCE • Audio/video tapes (or transcripts) • Written work • Teacher interviews/observations • Journal entries • Student comments on cover sheet • Student self-assessment	Number Sense – Whole No./Fractions (4) Number Relationships/No. Theory (8)
				Operations/Place Value (4) Operations (8)
				Estimation (4/8)
				Patterns/Relationships (4) Patterns/Functions (8)
				Algebra (8)
				Geometry/Spatial Sense (4/8)
				Measurement (4/8)
				Statistics/Probability (4/8)
				TASK CHARACTERISTICS
WHAT – OUTCOMES OF ACTIVITIES	LANGUAGE OF MATHEMATICS	MATHEMATICAL REPRESENTATIONS	CLARITY OF PRESENTATION	EMPOWERMENT COMMENTS
FINAL RATING 1 Solution without extensions 2 Solution with observations 3 Solution with connections or application(s) 4 Solution with synthesis, generalization, or abstraction	FINAL RATING 1 No or inappropriate use of mathematical language 2 Appropriate use of mathematical language some of the time 3 Appropriate use of mathematical language most of the time 4 Use of rich, precise, elegant, appropriate mathematical language	FINAL RATING 1 No use of mathematical representation(s) 2 Use of mathematical representation(s) 3 Accurate and appropriate use of mathematical representation(s) 4 Perceptive use of mathematical representation(s)	FINAL RATING 1 Unclear (e.g., disorganized, incomplete, lacking detail) 2 Some clear parts 3 Mostly clear 4 Clear (e.g., well organized, complete, detailed)	Motivation Flexibility Risk Taking Reflection Confidence Perseverance Curiosity/Interest Value Math

References

Chapter 1

National Council of Teachers of Mathematics. (1989). *Curriculum and evaluation standards for school mathematics.* Reston, VA: Author.

Romberg, T. A., Zarinnia, E. A., & Williams, S. R. (1989). *The influence of mandated testing on mathematics instruction: Grade 8 teachers' perceptions.* Madison: University of Wisconsin, National Center for Research in Mathematical Sciences Education.

Romberg, T. A., Zarinnia, E. A., & Williams, S. R. (1990). *Mandated school mathematics testing in the United States: A survey of state mathematics supervisors.* Madison: University of Wisconsin, National Center for Research in Mathematical Sciences Education.

Chapter 2

Apple, M. W. (1979). *Ideology and curriculum.* London: Routledge & Kegan Paul.

Ayres, L. P. (1918). History and present status of educational measurements. In S. C. Parker (Ed.), *The measurement of educational products: Seventeenth yearbook of the National Society for the Study of Education* (Part II). Bloomington, IL: Public School Publishing Co.

Begle, E. G., & Wilson, J. W. (1970). Evaluation of mathematics programs. In Begle, E. G. (Ed.), *Mathematics education,* 69th Yearbook of the NSSE (Part 1). Chicago: University of Chicago Press.

Bloom, B. S. (Ed.). (1956). *Taxonomy of educational objectives: The classification of educational goals. Handbook I: Cognitive domain.* New York: McKay.

Campbell, D. T., & Stanley, J. T. (1966). *Experimental and quasi-experimental designs for research.* Chicago: Rand McNally.

Clarke, D. J. (1987). The interactive monitoring of children's learning of mathematics. *For the Learning of Mathematics, 7*(1), 2–6.

Close, G., & Brown, M. (1988). *Graduated assessment in mathematics: Report of the SSCC study.* London: Department of Education and Science.

Collis, K. F. (1987). Levels of reasoning and the assessment of mathematical performance. In T. A. Romberg & D. M. Stewart (Eds.), *The monitoring of school mathematics: Background papers.* Madison: Wisconsin Center for Education Research.

Collis, K. F., Romberg, T. A., & Jurdak, M. E. (1986). A technique for assessing mathematical problem-solving ability. *Journal for Research in Mathematics Education, 17*(3), 206–221.

de Lange, J. (1987). *Mathematics, insight, and meaning: Teaching, learning and testing of mathematics for the life and social sciences.* Utrecht, The Netherlands: Rijksuniversiteit Utrecht.

Eash, M. J. (1985). Evaluation research and program evaluation: Retrospect and prospect: A reformulation of the role of the evaluator. *Educational Evaluation and Policy Analysis, 7*(3), 249–52.

Eisenberg, T. (1975). Behaviorism: The bane of school mathematics. *Journal of Mathematical Education, Science, and Technology, 6*(2), 163–71.

Eisner, E. (1976). Educational connoisseurship and criticism: Their form and function in educational evaluation. *Journal of Aesthetic Education, 10,* 173–79.

Fetterman, D. (1984). *Ethnography in educational evaluation.* Beverly Hills: Sage.

Foxman, D. D., Badger, M. E., Martini, R. M., & Mitchell, P. (1981). *Mathematical development secondary survey report no. 2.* London: Her Majesty's Stationery Office.

Foxman, D. D., Cresswell, M. J., Ward, M., Badger, M. E., Tucson, J. A., & Bloomfield, B. A. (1980). *Mathematical development primary survey report no. 1.* London: Her Majesty's Stationery Office.

Freeman, F. N. (1930). *Mental tests: Their history, principles and applications* (rev. ed.). Boston: Houghton Mifflin.

Gorth, W. P., Schriber, P. E., & O'Reilly, R. P. (1974). *Comprehensive achievement monitoring: A criterion-referenced evaluation system.* New York: Educational Technology Publishers.

Greene, H. A., Jorgensen, A. N., & Gerberich, J. R. (1953). *Measurement and evaluation in the elementary school* (2nd ed.). New York: Longmans.

Guba, E., & Lincoln, Y. (1981). *Effective evaluation.* San Francisco: Jossey-Bass.

Guralnik, P. B. (Ed.). (1985). *Webster's New World Dictionary.* New York: Prentice-Hall.

McLean, L. D. (1982). *Report of the 1981 field trials in English and mathematics: Intermediate division.* Toronto, Ontario: The Minister of Education.

National Coalition of Advocates for Students. (1985). *Barriers to excellence: Our children at risk.* Washington, DC: Author.

National Commission on Excellence in Education. (1983). *A nation at risk: The imperative for educational reform.* Washington, DC: US Government Printing Office.

National Council of Teachers of Mathematics. (1989). *Curriculum and evaluation standards for school mathematics.* Reston, VA: Author.

Odell, C. W. (1930). *Educational measurements in high school.* New York: Century.

O'Keefe, J. (1984). The impact of evaluation on federal education program policies. *Studies in Educational Evaluation, 10,* 612–74.

Patton, M. Q. (1980). *Qualitative evaluation methods.* Beverly Hills: Sage.

Peterson, P. L., Fennema, E., Carpenter, T. P., & Loef, M. (1989). Teachers' pedagogical content beliefs in mathematics. *Cognition and Instruction, 6*(1), 1–40.

Popkewitz, T. S. (1984). *Paradigm & ideology in educational research.* London: The Falmer Press.

Reinhard, D. (1972). *Methodology for input evaluation utilizing advocate and design teams.* Unpublished doctoral dissertation. The Ohio State University.

Romberg, T. A. (1987). *The domain knowledge strategy for mathematical assessment. Project Paper #1.* Madison, WI: National Center for Research in Mathematical Sciences Education.

Romberg, T. A. (1975). Answering the question—is "it" any good?—The role of evaluation in multi-cultural education through competency-based teacher education. In C. A. Grant (Ed.), *Sifting and winnowing: An exploration of the relationship between multi-cultural education and CBTE.* Madison, WI: Teacher Corps Associates.

Romberg, T. A. (1976). *Individually guided mathematics.* Reading, MA: Addison-Wesley.

Romberg, T. A. (1983). A common curriculum for mathematics. In G. D. Fenstermacher & J. I. Goodlad (Eds.), *Individual differences and the common curriculum.* Chicago: The University of Chicago Press.

Romberg, T. A. (Ed.). (1985). *Toward effective schooling.* New York: University Press of America.

Romberg, T. A., & Kilpatrick, J. (1969). Appendix D. Preliminary study on evaluation in mathematics education. In T. A. Romberg & J. W. Wilson (Eds.), *The development of tests.* NLSMA report no. 7. (pp. 281–98). Stanford, CA: School Mathematics Study Group.

Schoenfeld, A. H., & Herrmann, D. J. (1982). Problem perception and knowledge structure in expert and novice mathematical problem solvers. *Journal of Experimental Psychology: Learning, Memory, and Cognition, 8,* 484–94.

Scriven, M. (1974). Evaluation perspectives and procedures. In W. J. Popham (Ed.), *Evaluation in Education.* Berkeley: McCutchan.

Spearman, C. (1904). General intelligence objectively determined and measured. *American Journal of Psychology, 15*, 201–93.

Stake, R. E. (1974). *Program evaluation, particularly responsive evaluation.* Occasional Paper No. 5. Calamus: Western Michigan University Evaluation Center.

Stake, R. E., & Gjerde, C. (1974). An evaluation of the T-CITY. In Kraft et al. (Eds.), *Four evaluation examples: Anthropological, economic, narrative and portrayal.* AERA Monograph Series on Curriculum Evaluation, No. 7. Chicago: Rand McNally.

Swan, M. (1986). *The language of graphs: A collection of teaching materials.* Nottingham, UK: University of Nottingham, The Shell Centre for Mathematical Education.

Thorndike, E. L. (1904). *An introduction to the theory of mental and social measurements.* New York: Teachers College, Columbia University.

Tyler, R. W. (1931). A generalized technique for constructing achievement tests. *Educational Research Bulletin, 8,* 199–208.

Vergnaud, G. (1982). Cognitive and developmental psychology and research in mathematics education: Some theoretical and methodological issues. *For the Learning of Mathematics, 3*(2), 31–41.

Watson, G. (1938). The specific techniques of investigation: Testing intelligence, aptitudes, and personality. In G. M. Whipple (Ed.), *The scientific movement in education: Thirty-seventh yearbook of the National Society for the Study of Education* (Part II, pp. 365–66). Bloomington, IL: Public School Publishing.

Weinzweig, A. I., & Wilson, J. W. (1977). *Second IEA mathematics study: Suggested tables of specifications for the IEA mathematics tests. Working Paper I.* Wellington, New Zealand: IEA International Mathematics Committee.

Young, M. F. D. (1975). An approach to the study of curricula as socially organized knowledge. In M. Golby, J. Greenwald, & R. West (Eds.), *Curriculum design* (pp. 101–27). London: The Open University Press.

Chapter 3

Carpenter, T. P., & Fennema, E. (1988). *Research and cognitively guided instruction.* Madison, WI: National Center for Research in Mathematical Sciences Education.

Carpenter, T. P., Fennema, E., Peterson, P. L., & Carey, D. A. (1987). *Teachers' pedagogical content knowledge in mathematics.* Paper presented at the American Educational Research Association, Washington, DC.

Carpenter, T. P., Fennema, E., & Peterson, P. L. *Assessing children's thinking.* Unpublished working paper for the Cognitively Guided Instruction Project. Wisconsin Center for Education Research.

de Lange, J. (1987). *Mathematics, insight, and meaning: Teaching, learning and testing of mathematics for the life and social sciences.* Unpublished doctoral dissertation, Rijksuniversiteit Utrecht, The Netherlands.

Guilford, J. P. (1965). *Fundamental statistics in psychology and education.* New York: McGraw-Hill Book Company.

National Council of Teachers of Mathematics. (1989). *Curriculum and evaluation standards for school mathematics.* Reston, VA: Author.

Romberg, T. A. (1987). *The domain knowledge assessment strategy. Working Papers 87–1, School Mathematics Monitoring Center.* Madison: Wisconsin Center for Education Research.

Swan, M. (1985). *The language of functions and graphs.* UK: University of Nottingham, Shell Centre for Mathematical Education.

Vergnaud, G. (1982). Cognitive and developmental psychology and research in mathematics education: Some theoretical and methodological issues. *For the Learning of Mathematics, 3*(2), 31–41.

Chapter 4

Baron, J., Forgione, P., Rindone, D., Kruglenski, H., & Davey, B. (1989, March). *Toward a new generation of student outcome measures: Connecticut's Common Core of Learning assessment.* Paper presented at the annual meeting of the American Educational Research Association, San Francisco, CA.

California State Department of Education (1989). *A question of thinking: A first look at students' performance on open-ended questions in mathematics.* Sacrametno, CA: Author.

The California Achievement Test. (1985). Monterey: CTB/McGraw Hill.

The Comprehensive Test of Basic Skills. (1989). (4th ed). Monterey: CTB/McGraw Hill.

de Lange, J., van Reeuwijk, M., Burrill, G., & Romberg, T. A. (in press). *Learning and testing mathematics in context: The case: Data visualization.* National Council of Teachers of Mathematics: Reston, Va.

Edelman, J. F. (1980). The impact of the mandated testing program on classroom practices: Teacher perspectives. (Doctoral dissertation. University of California, Los Angeles) *Dissertation Abstracts International, 41,* 04.

The Iowa Test of Basic Skills. (1986). Chicago, IL: Riverside Publishing Co.

The Metropolitan Achievement Test. (1986). (6th ed). San Antonio, TX: The Psychological Corporation.

Meier, T. (1989, Jan/Feb). The case against standardized achievement tests. In *Rethinking Schools, Vol. 3*(2), 9–12.

Millman, J., Bishop, C. H., & Ebel, R. (1965). An analysis of test-wiseness. *Educational and Psychological Measurement, 25,* 707–26.

National Council of Teachers of Mathematics (1980). *An agenda for action: Recommendations for school mathematics of the 1980's.* Reston, VA: Author.

National Council of Teachers of Mathematics. (1989). *Curriculum and evaluation standards for school mathematics.* Reston, VA: Author.

Putnam, R. T., Lampert, M., & Peterson, P. L. (1989). Alternative perspectives on knowing mathematics in elementary schools. *Center for the learning and teaching of elementary school subjects.* Michigan State University, Lansing.

Romberg, T. A., Wilson, L., & Chavarria, S. (1990). *An examination of state and foreign tests.* Madison: Wisconsin Center for Education Research.

Romberg, T. A., Wilson, L., & Khaketla, M. (1989). *An examination of six standard mathematics tests for grade eight.* Madison: Wisconsin Center for Education Research.

Romberg, T. A., Zarinnia, A., & Williams, S. (1989). *The influence of mandated testing on mathematics instruction: Grade 8 teachers' perceptions.* Madison: University of Wisconsin, National Center for Research in Mathematical Sciences Education.

Science Research Associates Survey of Basic Skills. (1985). Chicago, IL: Author.

The Stanford Achievement Test. (1982). (7th ed). San Antonio, TX: The Psychological Corporation.

Stanley, J. C., & Hopkins, K. D. (1981). *Educational and psychological measurement and evaluation.* Englewood Cliffs, NJ: Prentice-Hall.

Chapter 5

Coley, R. J., & Goertz, M. E. (1990). *Educational standards in the 50 states.* Research Report 90–15. Princeton, NJ: Educational Testing Service.

Massachusetts Department of Education. (1987). *The 1987 Massachusetts Educational Assessment Program.* Quincy: Author.

National Council of Teachers of Mathematics. (1989). *Curriculum and evaluation standards for school mathematics.* Reston, VA: Author.

Shavelson, R. J. (1990). *Can indicator systems improve the effectiveness of mathematics and science education? The case of the U.S.* Santa Barbara: University of California.

National Assessment of Educational Progress (1988). *Mathematics objectives. 1990 Assessment: The nation's report card.* Princeton, NJ: Educational Testing Service.

Chapter 6

California State Department of Education (1985). *Mathematics framework for California public schools: Kindergarten through grade 12.* Sacramento: Author.

California State Department of Education (1987). *Mathematics model curriculum guide, K–8.* Sacramento: Author.

California State Department of Education (1989). *A question of thinking: A first look at student performance on open-ended questions in mathematics.* Sacramento: Author.

California State Department of Education (1989). *Survey of academic skills: Mathematics, grade 12.* Sacramento, CA: Author.

Cronbach, L. J. (1980). *Toward reform of program evaluation.* San Francisco: Jossey-Bass Publishers.

Honig, W. (1985). *Last chance for our children.* New York: Addison-Wesley.

Lester, F. K., Jr. (1978). Mathematical problem solving in the elementary school: Some educational and psychological considerations. In L. L. Hatfield & D. A. Bradbard (Eds.), *Mathematical problem solving: Papers from a research workshop.* Columbus, OH: ERIC Clearinghouse for Science, Mathematics, and Environmental Education.

Lester, F. K., Jr. (1982). Reflections about teaching mathematical problem solving in the elementary grades. In R. I. Charles & E. A. Silver (Eds.), *The teaching and assessing of mathematical problem-solving* (pp. 115–124). Hillsdale, NJ: Erlbaum.

Lesh, R. (1983, June). *Conceptual analyses of problem solving performance.* Paper presented at the Conference on Teaching Mathematical Problem Solving, San Diego State University, San Diego.

Lord, F. M. (1962). Estimating norms by item sampling. *Educational and Psychological Measurement, 22,* 259–67.

Mayer, E. (1983, June). *Implications of cognitive psychology for instruction in mathematical problem solving.* Paper presented at the Conference on Teaching Mathematical Problem Solving, San Diego State University, San Diego.

Millman, J. (1974). Criterion-referenced measurement. In W. J. Popham (Ed.), *Evaluation in Education* (309–397). Berkeley: McCutchan Publishing Company.

National Council of Teachers of Mathematics. *An agenda for action: Recommendations for school mathematics of the 1980s.* Reston, VA: NCTM, 1980.

Newell, A., & Simon, H. A. (1972). *Human problem solving.* Englewood Cliffs, NJ: Prentice-Hall.

Pandey, T. N. (1974). *Estimating the standard error of the mean in multiple matrix sampling when items are sampled with and without replacement.* Paper presented at the annual meeting of the American Educational Research Association, Chicago.

Pandey, T. N. (1983). *Structure for the assessment of problem solving.* Paper presented at the annual meeting of the American Educational Research Association, New Orleans.

Pandey, T. N., & Carlson, D. (1975). Assessing payoffs in the estimation of the mean using multiple matrix sampling designs. In D. N. M. De Gruijter & L. T. Th. van der Kamp (Eds.), *Advances in Psychological and Educational Measurement* (265–275). New York: Wiley.

Pandey, T. N., & Carlson, D. (1983). Application of item response models to reporting assessment date. In R. K. Hambleton (Ed.), *Applications of Item Response Theory* (212–229). Vancouver, BC: Educational Research Institute of British Columbia.

Polya, G. (1957). *How to solve it* (3rd ed.) Garden City, NJ: Doubleday.

Polya, G. (1965). *Mathematical discovery: On understanding, learning, and teaching problem solving* (Vol. 2). New York: Wiley.

Popham, W. J. (1973). *Evaluating instruction.* New Jersey: Prentice-Hall.

Resnick, L. B. (1983). Mathematics and science learning: a new conception. *Science, 220,* 477–78.

Schoenfeld, A. H. (1982). Some thoughts on problem solving research and mathematics education. In F. K. Lester & J. Garofalo (Eds.), *Mathematical Problem Solving: Issues in Research* (22–37). Philadelphia: The Franklin Institute Press.

Silver, E. A. (1982, January). Thinking about problem solving: Toward an understanding of meta-cognitive aspects of mathematical problem solving. Prepared for the Conference on Thinking, University of the South Pacific, Suva, Fiji.

Sternberg, R. J. (1981). Intelligence as thinking and learning skills. *Educational Leadership,* October, 1981.

Sternberg, R. J. (1983, February). Criteria for intellectual skills training. *Educational Researcher,* 6–12.

Chapter 7

College Board. (1985). *Academic preparation in mathematics: Teaching for transition from high school to college. New* York: Author.

Demana, F., & Waits, B. K. (1989). *Precalculus mathematics: A graphing approach.* Reading, MA: Addison-Wesley.

Demana, F., & Waits, B. (1990). Enhancing mathematics teaching and learning through technology. In T. Cooney (Ed.), *Teaching and learning mathematics in the 1990s* (pp. 212–222). Reston, VA: National Council of Teachers of Mathematics.

Demana, F. D., Foley, G., Harvey, J. G., Osborne, A., & Waits, B. K. *Results of the 1988–89 field test of precalculus: A graphing approach.* Unpublished manuscript.

Fey, J. T. (Ed.). (1984). *Computing and mathematics: The impact on secondary school curriculum.* Reston, VA: National Council of Teachers of Mathematics.

Fey, J. T. (1989). School algebra for the year 2000. In C. Kieran & S. Wagner (Eds.), *Research issues in learning and teaching algebra* (pp. 199–213). Reston, VA: National Council of Teachers of Mathematics.

Fey, J., & Heid, M. K. (1987, June). *Effects of computer-based curricula in algebra.* Paper presented at a meeting of National Science Foundation Project Directors. College Park, MD.

Goldenberg, E. P. (1988). Mathematics, metaphors, and human factors: Mathematical, technical, and pedagogical challenges in the educational use of graphical representation of functions. *Journal of Mathematical Behavior, 7,* 135–73.

Harvey, J. G. (1989). Placement test issues in calculator-based mathematics examinations. In J. W. Kenelly (Ed.), *The use of calculators in the standardized testing of mathematics* (pp. 25–46). New York: College Board & Mathematical Association of America.

Harvey, W., Schwartz, J., & Yerushalmy, M. (1988). *Visualizing algebra: The function analyzer* [Computer Software]. Pleasantville, NY: Sunburst.

Kaput, J. (1989). Linking representations in the symbol systems of algebra. In C. Kieran & S. Wagner (Eds.), *Research issues in learning and teaching algebra* (pp. 167–94). Reston, VA: National Council of Teachers of Mathematics.

Leinhardt, G., Zaslavasky, O., & Stein, M. K. (1990). Functions, graphs, and graphing: Tasks, learning and teaching. *Review of Educational Research, 60*(1), 1–64.

Lynch, J. K., Fischer, P., & Green, S. F. (1989). Teaching in a computer-intensive algebra curriculum. *Mathematics Teacher, 82,* 688–94.

National Council of Teachers of Mathematics. (1980). *An agenda for action: Recommendations for school mathematics of the 1980s.* Reston VA: National Council of Teachers of Mathematics: Author.

National Council of Teachers of Mathematics, Commission on Standards for School Mathematics. (1989). *Curriculum and evaluation standards for school mathematics.* Reston, VA: Author.

Rubenstein, R., Schultz, J., Hackworth, M., Flanders, J., Kissane, B., Aksoy, D., Brahos, D., Senk, S., & Usiskin, Z. (1988). *Functions, statistics and trigonometry with computers.* Chicago: University of Chicago School Mathematics Project.

Sarther, C., Hedges, L., & Stodolsky, S. *Formative evaluation of functions, statistics and trigonometry with computers.* Unpublished manuscript. University of Chicago School Mathematics Project, Chicago.

Senk, S. L. (1989). Toward algebra in the year 2000. In C. Kieran & S. Wagner (Eds.), *Research issues in learning and teaching algebra* (pp. 214–19). Reston, VA: National Council of Teachers of Mathematics.

Waits, B. K., & Demana, F. (1989). Computers and the rational root theorem—another view. *Mathematics Teacher, 82,* 124–25.

Waits, B. K., & Demana, F. (1988). *Master Grapher* [Computer software]. Reading, MA: Addison-Wesley.

Wiske, M. S., Zodhiates, P., Wilson, B., Gordon, M., Harvey, W., Krensky, L., Lord, B., Watt, M., & Williams, K. (1988, March). *How technology affects teaching.* Cambridge: Harvard Graduate School of Education, Educational Technology Center.

Chapter 8

Abo-Elkhair, M. E. (1980). An investigation of the effectiveness of using minicalculators to teach the basic concepts of average in the upper elementary grades. *Dissertation Abstracts International, 41,* 2980A. (University Microfilms No. 81–01, 953).

Bone, D. D. (1983). The development and evaluation of an introductory unit on circular functions and applications based on use of scientific calculators. *Dissertation Abstracts International, 44,* 1363A. (University Microfilms No. 83–22, 178).

Boyd, L. H., Lindquist, M. M., Harvey, J. G., & Waits, B. K. (1989). *Calculator-based arithmetic and skills test.* Washington, DC: Mathematical Association of America.

Casterlow, G., Jr. (1980). The effects of calculator instruction on the knowledge, skills, and attitudes of prospective elementary mathematics teachers. *Dissertation Abstracts International, 41,* 4319A. (University Microfilms No. 81–07, 547).

Cederberg, J., Demana, F. D., Harvey J. G., & Northcutt, R. A. (in press). *Calculator-based algebra test.* Washington, DC: Mathematical Association of America.

Colefield, R. P. (1985). The effect of the use of electronic calculators versus hand computation on achievement in computational skills and achievement in the problem-solving abilities of remedial middle school students in selected business mathematics topics. *Dissertation Abstracts International, 46,* 2168A. (University Microfilms No. 85–21, 950).

College Board. (1983). *Academic preparation for college: What students need to know and be able to do.* New York: Author.

Conference Board of the Mathematical Sciences. (1983). *New goals for mathematical sciences education.* Washington, DC: Author.

Connor, P. J. (1981). A calculator dependent trigonometry program and its effect on achievement in and attitude toward mathematics of eleventh and twelfth grade college bound students. *Dissertation Abstracts International, 42,* 2545A. (University Microfilms No. 81–24, 741).

Curtis, P. C., Jr., Harvey, J. G., Madison, B. L., & McCammon M. (in press). *Calculator-based basic algebra test.* Washington, DC: Mathematical Association of America.

Demana, F. D., & Leitzel, J. R. (1984). *Transition to college mathematics*. Reading, MA: Addison-Wesley.

Demana, F. D., Leitzel J. R., & Osborne, A. (1988). *Getting ready for algebra: Level 1 & Level 2*. Lexington, MA: D. C. Heath.

Demana, F. D., & Waits, B. K. (1990). *College algebra and trigonometry: A graphing approach*. Reading, MA: Addison-Wesley.

Demana, F. D., Waits, B. K., Foley, G. D., & Osborne, A. (1990). *College algebra and trigonometry: A graphing approach (Instructor's resource guide)*. Reading, MA: Addison-Wesley.

Elliott, J. W. (1980). The effect of using hand-held calculators on verbal problem solving ability of sixth-grade students. *Dissertation Abstracts International, 41,* 3464A. (University Microfilms No. 81–01, 829).

Epstein, M. G. (1968). Testing in mathematics: Why? what? how? *Arithmetic Teacher, 15*(4), 311–19.

Gimmestad, B. J. (1982). *The impact of the calculator on the content validity of Advanced Placement calculus problems*. Houghton, MI: Michigan Technological University, Department of Mathematical and Computer Sciences. (ERIC Document Reproduction Service No. ED 218 074).

Golden, C. K. (1982). The effect of the hand-held calculator on mathematics speed, accuracy, and motivation on secondary educable mentally retarded students (Grades 7–9). *Dissertation Abstracts International, 43,* 2311A. (University Microfilms No. 82–26, 927).

Harvey, J. G. (1989a). Placement test issues in calculator-based mathematics examinations. In J. W. Kenelly (Ed.), *The use of calculators in the standardized testing of mathematics*. New York: College Board & Mathematical Association of America, 25–33.

Harvey, J. G. (1989b). What about calculator-based placement tests? *The AMATYC Review, 11*(1, Part 2), 77–81.

Hopkins, B. L. (1978). The effect of a hand-held calculator curriculum in selected fundamentals of mathematics classes. (Doctoral dissertation, University of Texas at Austin, 1978). *Dissertation Abstracts International, 39,* 2801A.

————. The effect of a hand-held calculator curriculum in selected fundamentals of mathematics classes. In J. Lewis & H. D. Hoover, The effect of pupil performance on using hand-held calculators on standardized mathematics achievement tests. Paper presented at the Annual Meeting of the National Council on Measurement in Education, April 1981, Los Angeles.

Kenelly, J. W. (1989). *The use of calculators in the standardized testing of mathematics*. New York: College Board & Mathematical Association of America.

Kenelly, J. W., Harvey, J. G., Tucker T. W., & Zorn, P. (1990). *Calculator-based calculus readiness test*. Washington, DC: Mathematical Association of America.

Kilpatrick, J. (1985). *Academic preparation in mathematics: Teaching for transition from high school to college.* New York: College Board.

Kouba, V. L., & Swafford, J. O. (1989). Calculators. In M. M. Lindquist (Ed.), *Results from the fourth mathematics assessment of the National Assessment of Educational Progress* (94–105). Reston, VA: National Council of Teachers of Mathematics.

Leitzel, J. R., & Osborne, A. (1985). Mathematical alternatives for college preparatory students. In C. R. Hirsch & M. J. Zweng (Eds.), *The secondary school mathematics curriculum* (105–165). 1985 Yearbook of the National Council of Teachers of mathematics. Reston, VA: National Council of Teachers of Mathematics.

Leitzel, J. R., & Waits, B. K. (1989). The effects of calculator use on course tests and on statewide mathematics placement tests. In J. W. Kenelly (Ed.), *The use of calculators in the standardized testing of mathematics* (17–24). New York: College Board & Mathematical Association of America.

Lewis, J., & Hoover, H. D. (1981, April). *The effect on pupil performance of using hand-held calculators on standardized mathematics achievement tests.* Paper presented at the Annual Meeting of the National Council on Measurement in Education, Los Angeles.

Lindquist, M. M. (Ed.) (1989). *Results from the fourth mathematics assessment of the National Assessment of Educational Progress.* Reston, VA: National Council of Teachers of Mathematics.

Long, V. M., Reys, B., & Osterlind, S. J. (1989). Using calculators on achievement tests. *Mathematics Teacher, 82*(5), 318–25.

Mathematical Sciences Education Board. (1989). *Everybody counts: A report to the nation on the future of mathematics education.* Washington, DC: National Academy Press.

Mellon, J. A. (1985). Calculator based units in decimals and percents for seventh grade students. *Dissertation Abstracts International, 46,* 640A. (University Microfilms No. 85–10, 155).

Murphy, N. K. (1981). The effects of a calculator treatment on achievement and attitude toward problem solving in seventh grade mathematics. *Dissertation Abstracts International, 42,* 2008A. (University Microfilms No. 81–21, 439).

National Advisory Committee on Mathematical Education. (1975). *Overview and analysis of school mathematics grades K–12.* Washington, DC: Conference Board of the Mathematical Sciences.

National Assessment of Educational Progress. (1988). *Mathematics objectives: 1990 assessment.* Princeton: Educational Testing Service.

National Council of Teachers of Mathematics. (1980). *An agenda for action: Recommendations for school mathematics of the 1980s.* Reston, VA: Author.

National Council of Teachers of Mathematics. (1986, April). *Position statement: Calculators in the mathematics classroom.* Reston, VA: Author.

National Council of Teachers of Mathematics, Commission on Standards for School Mathematics. (1989). *Curriculum and evaluation standards for school mathematics.* Reston, VA: Author.

Romberg, T. A. (1984). *School mathematics: Options for the 1990s (Chairman's Report of a Conference).* Washington, DC: Department of Education, Office of the Assistant Secretary for Educational Research and Improvement.

Rule, R. L. (1980). The effect of hand held calculators on learning about: Functions, functional notation, graphing, function composition, and inverse functions. *Dissertation Abstracts International, 41,* 3866A. (University Microfilms No. 81–06, 048).

Chapter 9

Bell, R. C., & Hay, J. A. (1987). Differences and biases in English language examination formats. *British Journal of Educational Psychology, 57,* 212–20.

Biggs, J. B. (1976). Dimensions of study behaviour: Another look at ATI. *British Journal of Educational Psychology, 46,* 68–80.

Bolger, N. (1984, August). *Gender difference in academic achievement according to method of measurement.* Paper presented at 92nd annual convention of the American Psychological Association, Toronto, Ontario.

Choppin, B. (1975). Guessing the answer on objective tests. *British Journal of Educational Psychology, 45,* 206–13.

Crehan, K. D., Gross, L. J., Koehler, R. A., & Slakter, M. J. (1978). Developmental aspects of test-wiseness. *Educational Research Quarterly, 3*(1), 40–44.

Dwyer, C. A. (1979). The role of tests and their construction in producing apparent sex-related differences. In M. A. Wittig, & A. C. Petersen (Eds.). *Sex-related differences in cognitive functioning: Developmental issues* (pp. 335–53). New York: Academic Press.

Graf, R. G., & Riddell, J. C. (1972). Sex differences in problem solving as a function of problem context. *The Journal of Educational Research, 65*(10), 451–52.

Hambleton, R. K., & Traub, R. E. (1974). The effects of item order on text performance and stress. *The Journal of Experimental Education, 43*(1), 40–46.

Hill, K. T. (1984). Debilitating motivation and testing: A major educational problem—possible solutions and policy applications. *Research on Motivation in Education: Student Motivation, 1,* 245–74.

Kappy, K. A. (1980). Differential effects of decreased testing time on the verbal and quantitative aptitude scores of males and females. *Dissertation Abstracts International, 40*(12–A). 1980, Fordham University, Microfilm #DDJ82–18173.

Khampalikit, C. (1982). Race and sex differences in guessing behavior on a standardized achievement test in the elementary grades. *Dissertation Abstracts International, 43*(03–A). 1982, University of Pittsburgh. Microfilm #DDJ80–12788.

Kimball, M. M. (1989). A new perspective on women's math achievement. *Psychological Bulletin, 105*(2), 198–214.

Kleinke, D. J. (1980). Item order, response location, and examinee sex and handedness and performance on a multiple-choice test. *Journal of Educational Research, 73*, 225–29.

Klimko, I. P. (1984). Item arrangement, cognitive entry characteristics, sex, and text anxiety as predictors of achievement examination performance. *Journal of Experimental Education, 52*(4), 214–19.

Lane, D. S., Jr., Bull, K. S., Kundert, D. K., & Newman, D. L. (1987). The effects of knowledge of item arrangement, gender, and statistical and cognitive item difficulty on test performance. *Educational and Psychological Measurement, 47*, 865–79.

Leary, L. F., & Dorans, N. J. (1985). Implications for altering the context in which test items appear: A historical perspective on an immediate concern. *Review of Educational Research, 55*(3), 387–413.

Millman, J., Bishop, H., & Ebel, R. (1965). An analysis of test-wiseness. *Educational and Psychological Measurement, 25*, 707–26.

Murphy, R. J. L. (1982). Sex differences in objective test performance. *British Journal of Educational Psychology, 52*, 213–19.

Payne, B. D. (1984). The relationship of test anxiety and answer-changing behavior: An analysis by race and sex. *Measurement and Evaluation in Guidance, 16*(4), 205–10.

Plake, B. S., Ansorge, C. J., Parker, C. S., & Lowry, S. R. (1982). Effects of item arrangement, knowledge of arrangement, test anxiety and sex on test performance. *Journal of Educational Measurement, 19*(1), 49–57.

Plake, B. S., Patience, W. M., & Whitney, D. R. (1988). Differential item performance in mathematics achievement test items: Effects of item arrangement. *Educational and Psychological Measurement, 48*, 885–94.

Plass, J. A., & Hill, K. T. (1986). Children's achievement strategies and test performance: The role of time pressure, evaluation anxiety, and sex. *Developmental Psychology, 22*(1), 31–36.

Skinner, N. F. (1983). Switching answers on multiple-choice questions: Shrewdness or shibboleth? *Teaching of Psychology, 10*(4), 220–22.

Slakter, M. J. (1967). Risk taking on objective examinations. *American Educational Research Journal, 4*(1), 31–43.

Slakter, M. J., Koehler, R. A., & Hampton, S. H. (1970). Grade level, sex, and selected aspects of test-wiseness. *Journal of Educational Measurement, 7*(2), 119–22.

Speth, C. A. (1987). The effects of learning style, gender, and type of examination on expected test preparation strategies. (Doctoral dissertation, The University of Nebraska—Lincoln, 1987). *Dissertation Abstracts International, 49*(02–A).

Terwilliger, J. S. (1988). Item analysis of 1988 Twin Cities UMTYMP testing. Unpublished manuscript.

Watkins, D., & Hattie, J. (1981). The learning processes of Australian university students: Investigations of contextual and personological factors. *British Journal of Educational Psychology, 51*, 384–93.

Wild, C. L., Durso, R., & Rubin, D. B. (1982). Effect of increased test-taking time on test scores by ethnic group, years out of school, and sex. *Journal of Educational Measurement, 19*(1), 19–28.

Chapter 10

Baird, J. R., & Mitchell, I. J. (Eds.). (1986). *Improving the quality of teaching and learning: An Australian Case Study—the PEEL Project.* Melbourne, Victoria, Australia: Monash University Printery.

Biggs, J. (1988). The role of metacognition in enhancing learning. *Australian Journal of Education 32*(2), 127–38.

Clarke, D. J. (1985). *The IMPACT project: Project report.* Clayton, Victoria, Australia: Monash Centre for Mathematics Education.

Clarke, D. J. (1987). The interactive monitoring of children's learning of mathematics. *For the Learning of Mathematics, 7*(1), 2–6.

Clarke, D. J. (1989). *Assessment alternatives in mathematics.* A publication of the Mathematics Curriculum and Teaching Program (MCTP) Professional Development Project. Canberra, A.C.T., Australia: Curriculum Development Centre.

Clarke, D. J., Stephens, W. M. & Waywood, A. (1989). *Communication and the learning of mathematics: The Vaucluse Study—Supplement A.* Oakleigh, N.S.W., Australia: Australian Catholic University—Christ Campus.

Garofalo, J., & Lester, F. K. (1984). Metacognition, cognitive monitoring and mathematical performance. *Journal for Research in Mathematics Education 16*(3), 163–76.

Kilpatrick, J.(1985). Reflection and recursion. *Educational Studies in Mathematics 16*, 1–26.

Mason, J. (1984). Learning and doing mathematics. *Mathematics foundation course.* The Open University Press, UK.

National Council of Teachers of Mathematics. (1989). *Curriculum and evaluation standards for school mathematics.* Reston, VA : Author.

Rowe, M. B. (1978). Wait, wait, wait . . . *School Science and Mathematics, 78*(3), 207–16

Schoenfeld, A. (1985). *Mathematical problem solving.* London: Academic Press.

Stephens, W. M. (1982). *School work and mathematical knowledge.* Madison: Wisconsin Center for Educational Research.

Waywood, A. (1988). Mathematics and language: Reflections on students using mathematics journals. In R. Hunting (Ed.), *Language issues in learning and teaching mathematics*. Bundoora, N. S. W., Australia: Latrobe University.

White, R. T. (1986). Origins of PEEL. In J. R. Baird & I. J. Mitchell (Eds.), *Improving the quality of teaching and learning: An Australian Case Study—the PEEL Project*. Melbourne, Victoria, Australia: Monash University Printery.

Chapter 11

Biggs, J. B. & Collis, K. F. (1982). *Evaluating the quality of learning: The SOLO Taxonomy*. New York: Academic Press.

Bloom, B. S. (Ed.) (1956). *Taxonomy of educational objectives: The classification of educational goals. Handbook 1: Cognitive domain*. New York: Longman.

Bloom, B.S., Hastings, J. T., & Madaus, G. F. (1971). *Handbook on formative and summative evaluation of student learning*. New York: McGraw-Hill.

Chi, M. T. H., Feltovich, P. J., & Glaser, R. (1981). Categorization and representation of physics problems by experts and novices. *Cognitive Science*, 5(2), 121–52.

Collis, K. (1983). Development of a group test of mathematical understanding using superitem SOLO technique. *Journal of Science and Mathematics Education in South East Asia*, 6(1), 5–14.

Cureton, E. E. (1965). Reliability and validity: Basic assumptions and experimental designs. *Educational and Psychological Measurement*, 25(2), 326–46.

Dahlgren, L-O. (1984). Outcomes of learning. In F. Marton, D. Hounsell, & N. Entwistle (Eds.), *The experience of learning*. Edinburgh: Scottish Academic Press.

D'Ambrosio, U. (1979). Overall goals and objectives for mathematical education. In UNESCO International Commission on Mathematical Instruction, *New trends in mathematics teaching*. Paris: UNESCO.

Davis, P. J., & Hersh, R. (1981). *The mathematical experience*. Boston: Houghton Mifflin.

Freudenthal, H. (1983). Major problems in mathematics education. In M. Zweng, T. Green, J. Kilpatrick, H. Pollack, & M. Suydam (Eds.), *Proceedings of the Fourth International Congress on Mathematical Education*. Boston: Birkhauser.

Hambleton, R. K., & Swaminathan, H. (1985). *Item response theory: Principles and applications*. Boston: Kluwer-Nijhoff.

Johansson, B., Marton, F., & Svensson, L. (1985). An approach to describing learning as change between qualitatively different conceptions. In L. H. West & L. A. Pines (Eds.), *Cognitive structure and conceptual change*. Orlando: Academic Press.

Kulm, G. (Ed.). (1990). Assessing higher order thinking in mathematics. Washington, DC: American Association for the Advancement of Science.

Larkin, J. H. (1983). The role of problem representation in physics. In D. Gentner & A. Stevens (Eds.), *Mental models.* Hillsdale, NJ: Erlbaum.

Laurillard, D. (1984). Learning from problem solving. In F. Marton, D. Hounsell, & N. Entwistle (Eds.), *The experience of learning.* Edinburgh: Scottish American Press.

Leonard, F., & Sackur-Grisvald, C. (1981). Sur des regles implicites utilisees dans la comparison des nombres decimaux positifs [On two implicit rules used in comparison of positive decimal mumbers]. *Bulletin de l' APMEP, 327,* 47–60.

Marton, F. (1981). Phenomenography—describing conceptions of the world around us. *Instructional Science, 10*(2), 177–200.

Masters, G. N. (1982). A Rasch model for partial credit scoring. *Psychometrika, 47*(2), 149– 74.

Masters, G. N., & Wilson, M. (1988). *PC-CREDIT* [Computer Program]. Melbourne: University of Melbourne, Centre for the Study of Higher Education.

McCloskey, M. Caramazza, A., & Green, B. (1980). Curvilinear motion in the absence of external forces: Naive beliefs about motion of objects, *Science, 210,* 1139–41.

National Council of Teachers of Mathematics. (1980). *An agenda for action: Recommendations for school mathematics of the 1980's.* Reston, VA: Author.

Nesher, P. (1986). Learning mathematics: A cognitive perspective. *American Psychologist, 41*(10), 1114–22.

Nesher, P., & Peled, I. (1984). *The derivation of mal-rules in the process of learning.* Haifa, Israel: University of Haifa.

Resnick, L. B. (1982). Syntax and semantics in learning to subtract. In T. Carpenter, J. Moser, & T. A. Romberg (Eds.), *Addition and subtraction: A cognitive perspective.* Hillsdale, NJ: Earlbaum.

Resnick, L. B. (1984). Beyond error analysis: The role of understanding in elementary school arithmetic. In H. N. Cheek (Ed.), *Diagnostic and prescriptive mathematics: Issues, ideas and insights.* Kent, OH: Research Council for Diagnosis and Prescriptive Mathematics Research.

Romberg, T. (1983). A common curriculum for mathematics. In G. D. Festermacher & J. I. Goodlad (Eds.), *Individual differences and the common curriculum: Eighty-second yearbook of the National Society for the Study of Education.* Chicago: University of Chicago Press.

Romberg, T. A., Collis, K. F., Donovan, B. F., Buchanan, A. E., & Romberg, M. N. (1982). *The development of mathematical problem solving superitems.* (Report of NIE/EC Item Development Project.) Madison: Wisconsin Center for Educational Research.

Romberg, T. A., Jurdack, M. E., Collis, K. F. & Buchanan, A. E. (1982). *Construct validity of a set of mathematical superitems.* (Report of NIE/EC Item Development Project.) Madison: Wisconsin Center for Educational Research.

Saljo, R. (1984). Learning from reading. In F. Marton, D. Hounsell, & N. Entwistle (Eds.). *The experience of learning.* Edinburgh: Scottish Academic Press.

Samejima, F. (1969). Estimation of latent ability using a response pattern of graded scores. *Psychometrika, Monograph Supplement No. 17.*

Sandburg, J. A., & Barnard, Y. F. (1986). *Story problems are difficult, but why?* Paper presented at the annual meeting of the American Educational Research Association, San Francisco.

Schwartz, J. (1985). *The geometric supposer* [computer program]. Pleasantville, NY: Sunburst Communications.

Swan, M. (1983). *The meaning and use of decimals.* Nottingham, University of Nottingham: Shell Centre for Mathematical Education.

Van Hiele, P. M. (1986). *Structure and insight: A theory of mathematics education.* Orlando: Academic Press.

Vergnaud, G. (1983). Multiplicative structure. In R. Lesh & M. Landau (Eds.), *Acquisition of mathematics concepts and processes.* New York: Academic Press.

Von Glasersfeld, E. (1983). Learning as a constructive activity. In J. C. Bergeron & N. Herscovics (Eds.), *Proceedings of the fifth annual meeting of the PME-NA.* Montreal: Université de Montréal, Faculté des Sciences de l'Education.

Webb, N. L., Day, R., & Romberg, T. A. (1988). *Evaluation of the use of "Exploring Data" and "Exploring Probability."* Madison: Wisconsin Center for Education Research.

Wright, B. D., & Masters, G. N. (1982). *Rating scale analysis.* Chicago: MESA Press.

Wright, B. D., & Stone, M. (1979). *Best test design.* Chicago: MESA Press.

Wilson, M., & Iventosch, L. (1988). Using the Partial Credit model to investigate responses to structured subtests. *Applied Measurement in Education, 1*(4), 319–34.

Chapter 12

Becker, J. R., & Pence, B. J. (in press). The California case: Curriculum is what counts. In T. A. Romberg & E. A. Zarinnia (Eds.), *Four case studies on the impact of mandated testing.* Madison: Wisconsin Center for Education Research.

Bishop, A. J. (1988). *Mathematical enculturation.* Boston, MA: Kluwer Academic.

Bloom, B. S. (Ed.). (1956). *Taxonomy of educational objectives: The classification of educational goals. Handbook 1: Cognitive domain.* New York: Longman.

Brown, J. S., Collins, A., & Duguid, P. (1988). Situated cognition and the culture of learning. *Educational Researcher, 18*(1), 32–42.

California Assessment Program. (1987). *Survey of academic skills: Grade 12.* Sacramento: California State Department of Education.

California State Department of Education. (1985). *Mathematics framework for California public schools: Kindergarten through grade 12.* Sacramento: Author.

Cocking, R. R., & Mestre, J. P. (1988). *Linguistic and cultural influences on learning mathematics.* Hillsdale, NJ: Erlbaum.

Collis, K. F., Romberg, T. A., & Jurdak, M. E. (1986). A technique for assessing mathematical problem-solving ability. *Journal for Research in Mathematics Education, 17*(3), 206–21.

de Lange, J. (1987). *Mathematics, insight and meaning: Teaching, learning and testing of mathematics for the life and social sciences.* Utrecht, The Netherlands: Rijksuniversiteit Utrecht, Vakgroep Onderzoek Wiskundeonderwijs en Onderwijscomputercentrum.

Department of Education and Science. (1985). *Mathematics from 5 to 16.* London, UK: Her Majesty's Stationery Office.

Freudenthal, H. (1983). *Didactical phenomenology of mathematical structures.* Dordrecht, The Netherlands: D. Reidel.

Gale, D., & Shapley, L. S. (1967). College admissions and the stability of marriage. In M. S. Bell (Ed.), *Some uses of mathematics: A source book for teachers.* (Studies in Mathematics, Vol. XVI) Stanford, CA: School Mathematics Study Group.

GAIM Team. (1988). *Graded assessment in mathematics.* London, UK: Macmillan Education.

Mathematical Sciences Education Board. (1989). *Everybody counts: A report to the nation on the future of mathematics education.* Washington, DC: National Academy Press.

Mathematical Sciences Education Board. (1990). *On the shoulder of giants.* Washington, DC: National Academy Press.

Mellin-Olsen, S. (1987). *The politics of mathematics education.* Boston: D. Reidel.

National Assessment of Educational Progress. (1988). *Mathematics objectives: 1990 assessment.* Princeton: Educational Testing Service.

National Council of Teachers of Mathematics. (1989). *Curriculum and evaluation standards for school mathematics.* Reston, VA: Author.

Reich, R. B. (1983). *The next American frontier.* Harmondsworth, Middlesex, UK: Penguin Books.

Romberg, T. A. (1985, December). *The content validity for school mathematics in the US of the mathematics subscores and items for the Second International Mathematics Study.* Paper presented for the Committee on National Statistics, National Research Council of the National Academy of Sciences. Madison: Wisconsin Center for Education Research.

Romberg, T. A., & Kilpatrick, J. (1969). Appendix D. Preliminary study on evaluation in mathematics education. In T. A. Romberg & J. W. Wilson (Eds.), *The development of tests.* NLSMA report no. 7 (pp. 281–98). Stanford, CA: School Mathematics Study Group.

Romberg, T. A., & Wilson, J. W. (1969). *The development of tests.* NLSMA report, no. 7. Stanford, CA: School Mathematics Study Group.

Romberg, T. A., & Zarinnia, E. A. (Eds.). (in press a). *Four case studies on the impact of mandated testing.* Madison: Wisconsin Center for Education Research.

Romberg, T. A., & Zarinnia, E. A. (Eds.). (in press b). *A follow-up of the four case studies on the impact of mandated testing.* Madison: Wisconsin Center for Education Research.

Romberg, T. A., Zarinnia, E. A., & Williams, S. R. (1989). *The influence of mandated testing on mathematics instruction: Grade 8 teachers' perceptions.* Madison: University of Wisconsin, National Center for Research in Mathematical Sciences Education.

Romberg, T. A., Zarinnia, E. A., & Williams, S. R. (1990). *Mandated school mathematics testing in the United States: A survey of state mathematics supervisors.* Madison: University of Wisconsin, National Center for Research in Mathematical Sciences Education.

Rucker, R. (1987). *Mind tools: The five levels of mathematical reality.* Boston: Houghton Mifflin.

Scheffler, I. (1975, October). Basic mathematical skills: Some philosophical and practical remarks. In *The NIE Conference on basic mathematical skills and learning.* [Euclid, OH] *Volume I: Contributed position papers* (pp. 182–89). Los Alamitos, CA: SWRL Educational Research and Development.

School Mathematics Project. (1988). *SMP 11–16 OET handbook. (Trial ed. [rev.]).* Southampton, UK: Author.

Sirotnik, K. A. (1984). *An outcome-free conception of schooling: Implications for school-based inquiry and information systems.* Los Angeles: University of California, Center for the Study of Evaluation.

Stenmark, J. K. (1989). *Assessment alternatives in mathematics: An overview of assessment techniques for the future.* Prepared by EQUALS and the Assessment Committee of the California Mathematics Council, Campaign for Mathematics. Berkeley: University of California, Lawrence Hall of Science.

Westbury, I. (1980, January). Change and stability in the curriculum: An overview of the questions. In *Comparative studies of mathematics curricula: Changes and stability, 1960–1980 (pp.* 12–36). Proceedings of a conference jointly organized by the Institute for the Didactics of Mathematics (IDM) and the International Mathematics Committee of the Second International Mathematics Study of the International Association for the Evaluation of Educational Achievement (IEA). Bielefeld, FRG: Institut für Didaktik der Mathematik der Universität Bielefeld.

Wiggins, G. (1989, August). Teaching to the (authentic) test. *Educational Leadership 46*, 41–47.

Chapter 13

Clarkson, P. C. (1991). *Bilingualism and mathematics learning.* Geelong, Victoria, Australia: Deakin University Press.

Cuevas, G. J. (1984). Mathematics learning in English as a second language. *Journal for Research in Mathematics Education 15*(2), 134–44.

Leder, G. (1990). Teacher/student interactions in the mathematics classroom: A different perspective. In E. Fennema & G. Leder (Eds.), *Mathematics and gender.* New York: Teachers College Press.

Mathematical Sciences Education Board. (1989). *Everybody counts: A report to the nation on the future of mathematics education.* Washington, DC: National Academy Press.

Mathematical Sciences Education Board. (1990). *Reshaping school mathematics.* Washington, DC: National Academy Press.

National Council of Teachers of Mathematics. (1989). *Curriculum and evaluation standards for school mathematics.* Reston, VA: Author.

Newman, M. A. (1983). *The Newman language of mathematics kit.* Sydney, N. S. W., Australia: Harcourt Brace Jovanovich.

Pegg, J. E., & Davey, G. (1989). Clarifying level descriptors of children's understanding of some basic 2-D geometric shapes. *Mathematics Education Research Journal 1*(1): 16–27.

Secada, W. G. (1990). The challenge of a changing world for mathematics education. In T. Cooney & C. R. Hirsch, (Eds.), *Teaching and learning mathematics in the 1990s.* Reston, VA: National Council of Teachers of Mathematics.

Victoria Curriculum and Assessment Board. (1989). *Mathematics study design.* Melbourne, Victoria, Australia: Author.

Watson, I. (1980). Investigating errors of beginning mathematicians. *Educational Studies in Mathematics 11*, 319–30.

CONTRIBUTORS

James Braswell
Educational Testing Service
Princeton, New Jersey

Silvia Chavarria
Universidad de Costa Rica
San Jose

David Clarke
Australian Catholic University
(Christ Campus)
Oakleigh, Victoria, Australia

Philip C. Clarkson
Australian Catholic University
Ascot Vale, Victoria

John G. Harvey
University of Wisconsin-Madison

'Mamphono Khaketla
Ministry of Education
Lesotho

Margaret R. Meyer
University of Wisconsin-Madison

Tej Pandey
California Department of
Education
Sacramento

Sharon Senk
Michigan State University
East Lansing

Max Stephens
Ministry of Education
Victoria, Australia

Thomas A. Romberg
University of Wisconsin-Madison

Andrew Waywood
Vaucluse College
Richmond, Victoria, Australia

Norman Webb
University of Wisconsin-Madison

Linda Wilson
University of Wisconsin-Madison

Mark Wilson
University of California-Berkeley

E. Anne Zarinnia
University of Wisconsin-
Whitewater

AUTHOR INDEX

INDEX